Managing Technical Debt in Systems Engineering

Howard Kleinwaks

Managing Technical Debt in Systems Engineering

Springer

Howard Kleinwaks
Vienna, VA, USA

ISBN 978-3-032-30359-2 ISBN 978-3-032-30360-8 (eBook)
https://doi.org/10.1007/978-3-032-30360-8

This Springer imprint is published by the registered company Springer Nature Switzerland AG
The registered company address is: Gewerbestrasse 11, 6330 Cham, Switzerland

To Lee and the engineer you'll become

Preface

Like many others during the COVID-19 pandemic, I was spending a lot of time at home with my family, including my children, who were 3 and 4 years old. We filled the time as best we could with typical activities for children of that age: Paw Patrol action figures, board games, and daily long walks through the park. We improvised as best we could—we made up our own rules for Candy Land and made a life-sized Chutes and Ladders board on the driveway. Eventually, I wanted more, and I figured that since I wasn't going anywhere, why not go back to school and get another degree? I found the Colorado State University Doctor of Engineering program in Systems Engineering. It was completely online and therefore perfect for someone living in Virginia and not wanting to set foot in a public place during the pandemic.

As part of the program, I had to bring my own research topic to study. At the time, I was working on a schedule-constrained spiral development program. The cost and schedule were fixed for each spiral, so the only thing that could change if we ran into issues was performance. Given that each spiral was supposed to build upon the previous one, I was interested in understanding how to make the right decisions regarding performance reductions. How could we enable schedule success on the current spiral without negatively impacting the following spiral?

I thought the solution to this question was to develop a new requirements process for enduring architectures—develop a method to create requirements that span spirals and allow for partial fulfillment. I even used this idea as my initial research proposal.

As I thought about it over winter break, however, I realized that coming up with a new requirements process wasn't going to answer the question I was really interested in. New sets of requirements would still be subject to the same issues that I was already seeing. When time or money gets tight, the performance is going to suffer. How do we know if that performance degradation is acceptable, not just on the current spiral but on the future spirals? It was during this winter break that I read an article on technical debt.

Technical debt, this idea that decisions made to achieve a short-term benefit have long-term consequences, was exactly what I was looking for. Here was a way to capture the potential impacts of decisions on future capabilities. Sure, it was defined in software engineering, but surely someone had defined its applications in systems engineering, right? I could leverage that into a new decision-making framework for spiral development programs. I was excited, I had my approach to my dissertation. I emailed my advisors. All I had to do was find those systems engineering papers on technical debt, and I was off.

Except…there really weren't any systems engineering papers on technical debt. There was a lot written on technical debt in software engineering, but only some of it was applicable. The information at the time was scattered, and there wasn't a good, cohesive strategy on how technical debt should be used within a system development program. Like financial debt, not all technical debt is bad—it just needs to be properly managed. Yet, the discussions of management strategies were limited, and certainly nothing pertained to the complexities of modern system-of-systems development.

Therefore, I took on the challenge of defining and organizing the approach of technical debt within systems engineering as part of my doctoral research. I created an ontology of technical debt within systems engineering to standardize definitions and studied its impact on schedules. In my dissertation, I created a method to understand technical debt implications during decision-making to resolve current pressures while limiting the downstream impacts.

Leveraging my doctoral work, I started a technical debt management program at my job. We started looking at the decisions that we were making and their potential long-term consequences, both for the current spiral and for future spirals. We were able to identify which decisions would have larger and more costly impacts later so that we could mitigate those impacts. It turns out we were doing more than identifying technical debt; we were managing it.

It was this experience that inspired me to write this book. Technical debt within systems engineering projects is real, whether the projects use a traditional waterfall development cycle, a spiral development cycle, or an Agile approach. No project has infinite resources, and therefore compromises will always need to be made to release on time or stay within budget. These compromises often introduce technical debt into the system. This is not a bad thing! Technical debt can be advantageous, but it needs to be managed—we need to have "good debt" (like a mortgage paid monthly) and avoid "bad debt" (like a credit card bill that's never paid). To get to this point, we need to understand how to manage our technical debt.

This book provides just that—a process to identify and manage technical debt during systems development. Based on my experiences, I have provided a framework to start a technical debt management process in an organization to move the inevitable technical debt from "bad debt" to "good debt" to limit the risk of technical bankruptcy. Proper management of technical debt increases the likelihood of system success.

Vienna, VA, USA
2026

Dr. Howard Kleinwaks

Acknowledgments

Writing a book on technical debt certainly makes you think about whether each paragraph or page is good enough and whether or not you are making compromises in the work. It is challenging to convey the right level of information in a clear and concise form—not taking too many shortcuts and not being too long winded. I could not have done it without the help of several key individuals.

The idea for my Doctor of Engineering (D.Eng.) dissertation, which eventually became this book, started out as a discussion with Dr. Frank Turner on preventing decisions made today from limiting our options in the future on a spiral development program. I'm indebted to Frank not only for helping to form my initial ideas into coherent thoughts but also for reviewing an initial draft of this work. Frank also served on my doctoral committee at Colorado State University, along with my advisors Dr. Thomas Bradley and Ann Batchelor, and committee members Dr. Greg Marzolf and Dr. Daniel Wise. Their insights helped me refine several ideas about technical debt and its applications to systems engineering.

Thank you to my colleagues Matt Rich, Dr. Brian Kantsiper, Billy Palermo, Dean Ridgely, and Dr. Mike Sekerak for listening to me preach about technical debt and supporting me as I instituted a technical debt management program. Special thanks go out to Kaveh Kazemzadeh and Aaron Zucherman for helping with the actual day-to-day technical debt management activities and for providing insights as we turned a theoretical process into a practical one. Ben Bader and Dr. Alejandro Salado reviewed the draft and provided critical insights to make the book better and more readable. Mark Barrett helped me to garner support for my survey on technical debt among practicing systems engineers. Larry Gunn, Nate Fikru, Mike Butterfield, and Lt. Col Oesa Weaver all provided early commentary and reviews that helped me work through my initial thought processes.

Thank you to Michael Luby for deciding to publish my book and to Avanthika Babu, Nino Sponagel, and the entire editorial staff at Springer for shepherding this project to completion.

I acknowledge the use of generative AI, specifically Microsoft Copilot, to help write some of the Python code used to create the figures in this book.

Finally, an especially large thank you to my wife Suzanne and my children Ellie and Jake. During the writing of this book, I certainly had to make some compromises around family time. You were very patient with me as I first wrote a dissertation and then a book. Even though I tried to do most of the work while you were still asleep, there were still times when I was holed up in my office, and I really appreciate the time you all allowed me to have. I love you all.

Now, if you'll excuse me, I have to repay some of that technical debt I've built up with my family while you enjoy your upcoming read!

Competing Interests The author has no conflict of interest to declare that is relevant to the content of this book.

Ethics Approval No human or animal studies were performed in the preparation of this book.

Contents

About the Author

Dr. Howard Kleinwaks is the Chief Engineer of the National Security Division at Modern Technology Solutions, Inc (MTSI) and also the Tranche 1 Chief Engineer at the Space Development Agency. After graduating with Bachelor's and Master's degrees in Aerospace Engineering from MIT, he worked on several spiral development projects. This experience led him to ask the basic question, "How do we know that the decision we are making now will not have long-term impacts on our future spirals?" To answer this question, he completed a Doctor of Engineering degree at Colorado State University, studying the implications of technical debt on systems engineering development projects. Dr. Kleinwaks has over 20 years of experience working and managing complex systems development projects and has implemented a technical debt management program at SDA. He is a Certified Systems Engineering Professional (CSEP), Project Management Professional (PMP), and a Professional Scrum Master, and his experience stretching across systems engineering, project management, and Agile development informs the development of this book for like-minded readers.

Part I
Introduction to Technical Debt

Chapter 1
Introduction

Abstract Systems engineering projects often reach the stage where the developer feels that everything is taking longer than it should or where they realize that an earlier shortcut is creating longer-term problems. This phenomenon is called technical debt. While a familiar part of the lexicon in software engineering, technical debt is emerging as a concern within systems engineering. Technical debt within systems engineering is the quantitative impact on the long-term health of the system accrued as a result of a technical compromise made to achieve a short-term benefit. While related to risks, technical debt accrual follows a different process that must be managed separately.

Keywords Technical debt summary · Technical debt impacts · Comparison of techincal debt and risk · Prevalence of techincal debt in systems engineering

Have you ever gotten to a point in a project where everything is taking longer than it should? Or requires more effort? Or costs more? Can you trace these issues back to a decision that seemed like a good idea at the time but turns out it really wasn't? If so, you are likely experiencing the effects of technical debt. And you're not alone. A recent Wall Street Journal report [1] blames technical debt for issues like the cancellation of 13,000 Southwest airline flights during the 2022 holiday season and high-profile cyber-attacks, and puts the cost of addressing technical debt across multiple industries at $1.52 trillion (yes, that's trillion with a T).

So, what is technical debt? Technical debt is the long-term negative consequence of making a technical compromise in a project to achieve a short-term benefit, such as meeting a release date or reducing immediate costs. Anytime you take a shortcut or defer "the right solution" because you "don't have time for it" or "it is too expensive," you are taking on technical debt. By deferring the work, you are "borrowing" time or money (the amount saved by not doing the work now) with a promise to "pay it back" later (by fixing the problem at a later date). The amount saved is the principal on the loan. However, like financial loans, technical debt also accrues interest—it may take more time or effort to fix the problem in the future than it would to fix it now. If left

H. Kleinwaks, *Managing Technical Debt in Systems Engineering*,
https://doi.org/10.1007/978-3-032-30360-8_1

unchecked, too much interest accumulates, and too many problems remain unfixed, eventually making the system unusable until the debt is repaid.

The concept of technical debt originated in software engineering to explain the need to refactor (clean up) a code base to management. Refactoring code is transparent to the user but significantly helps the developers. Given the lack of a direct return on investment from the user's viewpoint, management may not see the need to take this step. However, forcing developers to use an unrefactored code base means that every time the code is used, it takes the developers longer than it should, increasing costs and delaying delivery. These delays and cost increases are the interest on the technical debt and decrease the overall system value. Like financial debt, too much technical debt is not healthy for a system.

While much of the research on technical debt is focused on software, technical debt itself is present in all engineering disciplines, especially as these disciplines move toward more iterative design practices with emphases on delivering minimum viable products (MVP) and speed to market. Modern systems are incredibly complex and interconnected cyber-physical systems, and what appears to be a small technical concession in one component can reverberate throughout the rest of the system. The rise of agile systems engineering methods and incremental development creates schedule constraints, exerts pressure on the development of cyber-physical systems, and establishes an environment ripe for the creation of technical debt.

The result: increased development times and cost, decreased performance, or decreased usability, three things that can significantly impact the success of a system. This book provides techniques to manage technical debt within systems engineering and to mitigate its impact. Systems engineers are in a unique position within system development, as they are responsible for considering both the business and technical aspects in the realization of the product [2] and fully understand the impacts that a change in one area has on the rest of the system. Therefore, systems engineers must be aware of technical debt and how it can flow across component boundaries to impact the final product.

My own experiences with technical debt while working on multiple systems using spiral development led me to question whether or not there was substantial research on technical debt focused on systems engineering. To my surprise, I found out that there wasn't much! This result led me to pursue my Doctor of Engineering degree in Systems Engineering, focusing on the presence of technical debt in systems engineering. Much of this book is based on my research.

In this book, we will discuss technical debt with a focus on its applications to systems development and systems engineering. We'll explore the origins of technical debt and develop a systems-engineering-focused lexicon that will enable us to define how and when technical debt appears within the systems engineering lifecycle. Finally, we'll discuss technical debt management strategies and how implementing these strategies can reduce the risk of technical bankruptcy. We'll review case studies and examples throughout the book to show how technical debt has impacted real systems with real consequences.

1.1 What is Technical Debt?

We start with the following definition of technical debt in systems engineering:

Definition Technical debt is a metaphor reflecting technical compromises that can yield short-term benefit but may hurt the long-term health of a system [3].

The metaphorical aspect of technical debt is the financial similarity, suggesting that the technical problems resemble financial problems in terms of borrowing, repayment, and consequences. However, to measure, manage, and control technical debt, we must define it as something tangible instead of a metaphorical reference. Therefore, in this book, we'll use the following definition of technical debt when referring to its quantifiable impact on a system:

Definition Technical debt is the quantitative impact on the long-term health of the system accrued as the result of a technical compromise made to achieve a short-term benefit [4].

Parsing this definition shows three distinct components: a technical compromise, a short-term benefit, and a long-term impact on the health of the system, which is typically interpreted as a negative impact. A technical compromise refers to a decision that reduces the performance of the system in one area to achieve a benefit in another area (cost, schedule, or yes, performance). For something to be technical debt, and not just a bad decision, there needs to be a benefit as a result of the compromise. This benefit can be a cost savings, a faster release, or an increase in performance in another area of the system. These compromises are usually made in response to the needs of a stakeholder who is concerned about an immediate issue. Meeting these time-critical needs is sometimes referred to as "fighting the alligator closest to the boat." At face value, this seems like a good idea. After all, who wants hungry man-eating alligators in the boat?

The critical aspect of technical debt, however, is that these compromises can carry negative long-term impacts. You successfully kept the alligator from climbing into the boat by pushing it away with the oar, but broke the oar in the process. Now, how are you going to steer around the rocks? Quick reactions to pressing needs have a tendency to only consider the near-term implications of the decision (no alligator) and not the longer-term impacts (avoiding the rocks).

1.1.1 Is All Technical Debt Bad?

Like financial debt, not all technical debt is bad—there are often very good reasons for making technical compromises: being first to market, meeting the schedule, reducing costs, and keeping stakeholders happy are all positive outcomes. However, you need to be aware of the potential for long-term consequences and consider them when making decisions. In the alligator example, if the fishing pole were used instead of the oar, then we'd still be able to steer around the rocks! We wouldn't be able to fish, but we'd make it out of the river (and more importantly, we would have the short-term benefit of being alive).

Technical debt is not necessarily due to poor choices or bad decisions. For many reasons, the short-term benefit may be worth making the technical compromises. The key is to manage the ensuing technical debt to prevent it from having a long-term impact, which is what this book is about.

1.1.2 How Does Technical Debt Enter a System?

Sometimes we are aware that we are making a technical compromise when it is made—choosing to skip a full peer review, selecting not to perform a specific test, or using inferior parts are all conscious decisions that are made to achieve other benefits such as preventing schedule delays and saving budget. However, not all of the compromises which give rise to technical debt are made intentionally. Unintentional decisions create a significant amount of technical debt. For example, insufficient requirements, part-level optimization that negatively impacts the rest of the system, and inferior documentation are all potential sources of technical debt that may not be the result of an intentional compromise. The existence of these sources of technical debt requires the system developer to be proactive in looking for and measuring the impact of these hidden technical compromises.

1.1.3 What Are the Impacts of Technical Debt?

Once technical debt is in the system, it can impact both system development and system utility. With large amounts of technical debt, system development slows down. Developers take more time to understand the current implementation and to correct problems. Sometimes, they are forced to execute a complete redesign. Once released, the system may not perform as advertised, which can decrease the user experience and reduce the likelihood of system acceptance. Just as unpaid financial debt creates significant consequences (bad credit rating, increased amounts owed, bankruptcy, repossession, etc.), unpaid technical debt creates consequences (increased costs, delayed releases, decreased performance) that impact system success.

One of the most pernicious qualities of technical debt is that it compounds—technical debt in one area of the system can create impacts in other areas of the system. As an example, consider the impacts of technical debt on a project schedule. Traditional schedule risk assessments analyze a schedule to provide an estimated completion date by applying probability distributions to task lengths. These assessments assume that a task is completed "perfectly" and does not hinder the following task. However, technical debt breaks this assumption. A task may complete on time, but if it accrues significant technical debt, then following tasks, either immediate successors or significantly farther downstream, may take substantially longer than planned. This effect

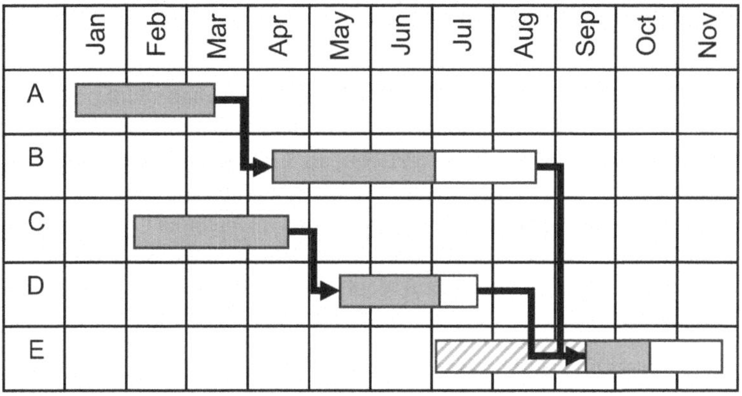

Fig. 1.1 Effect of technical debt on task duration

can be seen in Fig. 1.1, which shows a simple Gantt chart schedule. The white bars indicate the delay in the completion of tasks B, D, and E due to technical debt incurred during the development of tasks A and C—note that the durations of A and C are unchanged! This technical debt not only delays the start of task E (shown as the patterned bar), but it also compounds to extend the duration of task E. The technical debt on the initial tasks delays the completion of the entire project, without delaying the tasks where the compromises are made. Of course, technical debt is not the only reason why a project may be late, and there are many project management resources that discuss how to manage the schedule. However, unmanaged technical debt is a time bomb waiting to go off that can have serious implications for a schedule which may not be caught by traditional techniques. Therefore, it is critical that technical debt is tracked and managed to mitigate the downstream effects.

1.2 A Quick Case Study

Let's return to the winter of 2022 when Southwest Airlines cancelled all those flights. How did technical debt play a role in Southwest experiencing greater disruptions than its competitors? Simply put, Southwest decided to forgo upgrades to its computer system for years (the technical compromise). While details are not clear, we can assume that Southwest did not want to spend the money for the upgrades (the short-term benefit of increased profits through lower costs). However, failing to upgrade the systems puts Southwest at greater risk for larger disruptions (the negative impact on the long-term health of the system). Instead of routinely updating their systems, which would cost the airline money, Southwest delayed the updates until they experienced a catastrophic failure. In the end, Southwest cancelled 17 times as many flights as Delta Airlines and faced a cost of over $800 million [5]. We'll explore this example in more detail in Sect. 5.1.

1.3 Technical Debt Versus Risks

But isn't technical debt just a risk? Won't a robust risk management process handle technical debt? After all, risk management requires identifying unfavorable events and determining the likelihood and consequences of its occurrence. A strong risk management program takes active steps to mitigate both the likelihood and consequences of these unfavorable events and maintains management reserve in case the risk is realized. Shouldn't this process be able to handle technical debt?

The short answer is no: technical debt and risks have a simple but highly consequential difference:

- Risks are ***uncertain events*** that, if they occur, ***have a negative effect*** on the system [6]
- Technical debt occurs because of an ***event which occurred*** (the technical concession), but the ***magnitude of the effect is uncertain***

Risks are generally caused by events outside the control of the systems engineer. Risk management looks to prevent these events from happening or to minimize the known consequences of the event. Technical debt, however, occurs based on a compromise made by the systems engineer (or their team). There may be external pressure that leads to the compromise, and the compromise may have been unintentional; however, a technical compromise was still made. Therefore, the technical debt can be identified at its point of insertion into the system. The outcome of technical debt, however, may be less certain, as the total impact may depend on *when*, *how*, and *if* the debt is repaid.

Therefore, technical debt needs a different management process from risk management, although technical debt can arise from risk acceptance. The correlation between managing technical debt and risk management is discussed in more detail in Chap. 7.

1.4 Prevalence of Technical Debt in Systems Engineering

While we've provided an overview of technical debt so far, there is still the outstanding question—is technical debt a problem within systems engineering processes? After all, systems engineering has been around for a long time and has a long history of developing different management processes.

We could rely on personal experience to support the notion that technical debt occurs within systems engineering projects. Most systems engineers and system developers have asked themselves questions like "why is this taking so much longer than I thought it would?" and made statements like "knowing what I know now, I would have done this differently" during system development. They experience changes throughout the system lifecycle, due to requirements changes, test failures, and inadequacies in the design. Are these changes the result of technical debt? Does

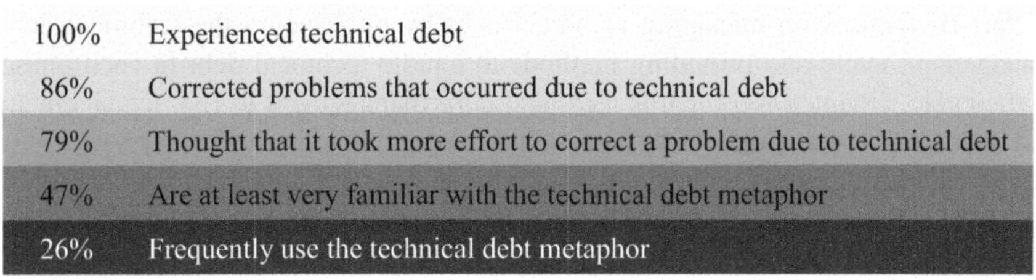

Fig. 1.2 Summary of technical debt survey results [8]

technical debt occur in systems engineering projects with our defined lifecycles and review paths? Or have we avoided the problem through our processes and techniques?

Instead of personal experience, we'll use the results of a 2023 survey of practicing systems engineers [7] which examined whether or not technical debt occurred within systems engineering. The results of the survey, shown in Fig. 1.2, clearly indicate that while technical debt is common in systems engineering, it is not commonly discussed among systems engineers. 100% (!) of the engineers surveyed experienced the effects of technical debt, but less than half of them used consistent terminology, making communication and problem-solving difficult.

The results of this survey highlight the need for this book—technical debt is real in systems engineering projects, but there are no existing processes to manage it. This book addresses this gap by providing a common ontology for technical debt, including definitions of its causes and types, and a management process to help limit the impacts of technical debt buildup in complex projects.

1.5 Outline of This Book

Part I of this book provides a short history of the concept of technical debt, including its origins in software engineering. Chapter 2 provides the origins of technical debt and shows how it translated from software engineering to systems engineering. Chapter 3 introduces key concepts for managing technical debt within systems engineering, including an ontology for technical debt, a technical debt concept map, and a discussion of what is and is not technical debt.

Part II focuses on the causes and types of technical debt and how they manifest within the systems engineering lifecycle. Chapter 4 identifies the different causes of technical debt, focusing on schedule pressure, cost pressure, and other factors that drive a system developer to make a technical compromise. Chapter 5 introduces the different types of technical debt found within systems engineering and provides short examples of each one from real-world scenarios. Chapter 6 discusses how technical debt appears throughout the systems engineering lifecycle and its impact on project schedules and system acceptance.

Part III focuses on managing technical debt by introducing the technical debt management cycle and providing methods to handle technical debt in each phase of the cycle. Chapter 7 compares technical debt management to risk management and introduces the technical debt register as a management tool. Chapter 8 provides ways to identify existing technical debt and to predict new technical debt as part of the project management cycle. Chapter 9 focuses on measuring technical debt by diving deeper into the concepts of principal, interest, taxes, and fees and providing methods to quantify the impacts of technical compromises. Chapter 10 discusses how to prioritize and repay technical debt, and Chapter 11 provides methods to continuously monitor the system. Chapter 12 discusses technical bankruptcy: the state where the technical debt has accumulated to the point where the project grinds to a halt until some or all of the debt is repaid. Chapter 13 concludes the book and provides the first steps to take to implement a technical debt management program in your organization.

References

1. Mims, C. (2024, March 1). The Invisible $1.52 Trillion Problem: Clunky Old Software. *The Wall Street Journal.* https://www.wsj.com/tech/personal-tech/the-invisible-1-52-trillion-problem-clunky-old-software-f5cbba27
2. Walden, D. D., Roedler, G. J., Forsberg, K. J., Hamelin, R. D., & Shortell, T. M. (Eds.). (2015). *Systems Engineering Handbook* (4th ed.). John Wiley and Sons.
3. Kleinwaks, H., Batchelor, A., & Bradley, T. H. (2023). Technical Debt in Systems Engineering - A Systematic Literature Review. *Systems Engineering, 26*(5), 675-687. https://doi.org/10.1002/sys.21681
4. Kleinwaks, H., Batchelor, A., & Bradley, T. H. (2023, September). An Ontology for Technical Debt in Systems Engineering. *IEEE Open Journal of Systems Engineering, 1*, 111-122. https://doi.org/10.1109/OJSE.2023.3316395
5. Dooley, R. (2023, January 6). *Southwest's $825 Million Loss Signals Why Companies Shouldn't Put Off Internal Investments.* Forbes. https://www.forbes.com/sites/rogerdooley/2023/01/06/southwest-meltdown-could-cost-825-million/
6. Project Management Institute, Inc. (2017). *A Guide to the Project Management Body of Knowledge (PMBOK Guide)* (6th ed.). Project Management Institute, Inc.
7. Kleinwaks, H., Batchelor, A., & Bradley, T. H. (2023). An Empirical Survey on the Prevalence of Technical Debt in Systems Engineering. *INCOSE International Symposium, 33*(1), 1640-1658. https://doi.org/10.1002/iis2.13104
8. Kleinwaks, H. (2023). An Empirical Survey on the Prevalence of Technical Debt in Systems Engineering [Poster Presentation]. *INCOSE International Symposium.* Honolulu, HI, United States.

Chapter 2
What is Technical Debt?

Abstract The metaphor of technical debt originated in the field of software engineering and is based on a financial analogy. Software development techniques, such as Agile development, that encourage speed of delivery also increase the risk of taking shortcuts to meet schedules and thereby introduce technical debt. The research on technical debt within software engineering has exponentially increased since 2010. The research on technical debt within systems engineering is more limited. However, systems engineering and software engineering follow similar lifecycles and approaches, and systems engineering is moving to more Agile development processes. This shift, combined with the increasing reliance of systems on software, means that technical debt within systems engineering projects must be managed.

Keywords Origins of technical debt · Software engineering versus systems engineering · Agile systems engineering · Technical debt · Agile development

The term "technical debt" has entered the public domain and is discussed by all levels of organizations, from junior developers up to executives. However, there is no widely accepted, common understanding of technical debt. Technical debt is used to refer to incomplete work, errors, schedule delays, or the potential for unrealized negative consequences. Without a consistent definition, technical debt means different things to different organizations, allowing some organizations to consider technical debt to be only a "software" or "IT" problem. Instead, technical debt is a problem across all fields and especially in systems engineering. In this chapter, we review the history of the technical debt metaphor and define what should and should not be considered technical debt, especially as it applies to systems engineering.

H. Kleinwaks, *Managing Technical Debt in Systems Engineering*,
https://doi.org/10.1007/978-3-032-30360-8_2

2.1 Origins of the Technical Debt Metaphor

Software developer Ward Cunningham is credited with the first use of the term technical debt, which he used to describe the need to refactor software to his management, who were not well versed in the intricacies of software development. In software, changing requirements, new developers, and bug fixes combine to clutter the code and eventually may make it unusable. To fix this problem, software developers need to refactor, or clean up, the code through optimization and consolidation of multiple functions. Refactoring code is not a sign of poor development; rather, it is a consequence of the inherent complexity of modern software. Cunningham, who was developing a financial application, decided to describe the need to refactor code by using an analogy to something his management would understand: money.

> If we failed to make our program align with what we then understood to be the proper way to think about our financial objects, then we were going to continue to stumble on that disagreement which is like paying interest on a loan [1].

In this quote, he is stating that having code that is misaligned with the objectives, whether due to bad code or changing requirements, will cost additional development time through "stumbling on disagreements." If the code doesn't align with the correct viewpoint, then the team would spend more time arguing over how to fix the code than it would take to actually fix it. This extra time is the interest on the loan: the development team spends extra effort to accomplish the same goals, just like financial interest is extra money spent on the initial loan.

At the 1992 Object-Oriented Programming, Systems, Languages and Applications (OOPSLA) conference, Cunningham elaborated on his metaphor in a published report, stating:

> Although immature code may work fine and be completely acceptable to the customer, excess quantities will make a program unmasterable, leading to extreme specialization of programmers and finally an inflexible product. Shipping first time code is like going into debt. A little debt speeds development so long as it is paid back promptly with a rewrite. Objects make the cost of this transaction tolerable. The danger occurs when the debt is not repaid. Every minute spent on not-quite-right code counts as interest on that debt. Entire engineering organizations can be brought to a stand-still under the debt load of an unconsolidated implementation, object- oriented or otherwise [2].

In this quote, "not-quite-right code" indicates software that has some form of compromise: it doesn't completely work, it needs to be refactored, or it contains other elements that make it "not-quite-right" for the task it is supposed to perform [1].

Cunningham later elaborated on the analogy through further interviews to expand how technical debt occurs and why the financial analogy makes sense. The general concept is that the development team borrows productivity from the future to get a product released (the loan). By taking a shortcut and releasing "not-quite-right" code, the development team has increased its productivity (demonstrated through the release) at the potential expense of future productivity. The future productivity is the interest on the amount borrowed, and occurs when future development efforts take longer (and are therefore less productive) than they otherwise should have been. Cunningham also makes the crucial point that technical debt, like financial debt, is not necessarily a bad thing, as long as there is a plan to pay it off [1]. Since Cunningham first published this concept in 1992, the application and usage of metaphor have significantly expanded.

As recognition of the software problems identified with technical debt grew, using technical debt only as a metaphor was not sufficient. The problems behind the metaphor were real. These problems affected the ability to both develop software and deliver capability to the users. The metaphor put a name to the problem; however, to resolve the problem, technical debt needs to move from the conceptual realm to the measurable realm and from metaphor to a tangible object. If technical debt can be measured, then it can be managed and controlled. Before it can be measured, the definition needs to be refined and the terminology standardized. In the next few sections, we'll review the financial metaphor and early published works that introduced several key concepts in technical debt and which provide the foundation for building our definition of technical debt within systems engineering.

2.1.1 *The Financial Metaphor*

Cunningham used a financial metaphor for technical debt since both developers and managers understood the concepts. Since then, the use of financial terms to define technical debt has increased. The following financial terms are essential to understanding technical debt [3, 4]:

- Debt: an amount owed by one party to another
- Principal: the original amount borrowed
- Interest: additional amount, beyond principal, owed by the borrower. Interest is the cost of borrowing the principal and may be a fixed amount or determined by an interest rate
- Interest rate: the percentage of the principal owed to the borrower, used to calculate the interest
- Compound interest: Debt is often paid back over time in scheduled installments. The amount owed with each installment is based on the current principal and the interest rate. Interest compounds when payments are not made, and therefore the interest rate is applied to both the principal and the previously owed interest.
- Fees: an expense for using a service. Fees are due every time the service is used and do not vary with interest rate or the current loan balance.
- Taxes: required levies on an organization by an external entity, typically a governmental body

Figure 2.1 shows the impact of unmanaged financial debt, exhibited by missing payments. Missing payments increase both the total amount paid (solid line) and the number of months to repay the loan (dashed line). Similarly, unmanaged technical debt increases the cost to the system developer and the development schedule.

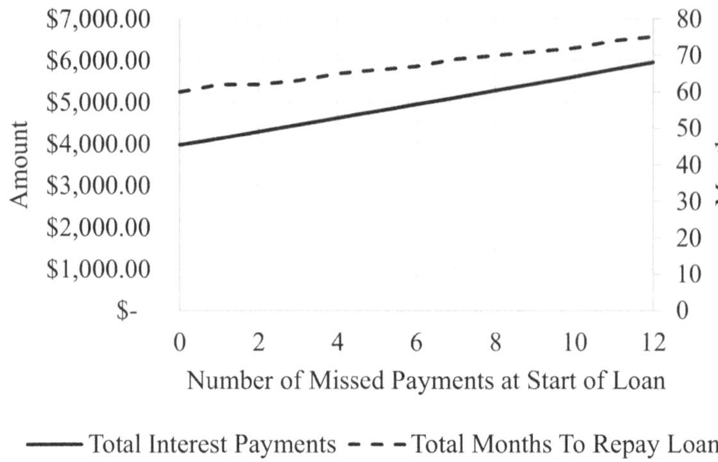

Fig. 2.1 Effect of missing interest payments on loan payoff

2.1.2 Initial Public Works and the Technical Debt Quadrant

In 2008, Steve McConnell published a white paper on managing technical debt [5]. He identified items that are not technical debt, specifically calling out that not all incomplete work is debt. Technical debt is only something that requires "interest payments;" it introduces additional work or decreases future productivity. This white paper started the standardization of the technical debt terminology.

In October of 2009, Martin Fowler created a blog post titled "Technical Debt Quadrant" [6]. In this post, Fowler argues that it is important to understand how technical debt enters a system. If we understand how technical debt occurs, then we can implement appropriate management strategies. He qualified technical debt as either reckless or prudent, where reckless debt is introduced into the system without a management plan and prudent debt is accompanied by a management plan. He also qualified the technical debt as either inadvertent or deliberate, where inadvertent debt is a byproduct of the development process while deliberate debt is the result of known action on the part of the system developer. A modified version of Fowler's technical debt quadrant is shown in Fig. 2.2. In the quadrant, it is more desirable to move from left to right and bottom to top as indicated by the direction of the arrows.

Within the quadrant, four specific types of technical debt are defined: negligent, strategic, unintentional, and tactical. Negligent debt occurs when system developers recklessly and deliberately skip steps or intentionally make technical compromises without a proper management strategy. This is a case of "knowing what you are doing is wrong, but doing it anyway." Strategic debt occurs when the debt accrual is deliberate and prudent, such as taking a short-cut to meet a release date (deliberate) while having a plan in place to correct it later (prudent). This category of technical debt can be beneficial to the development team and is similar to taking out a mortgage

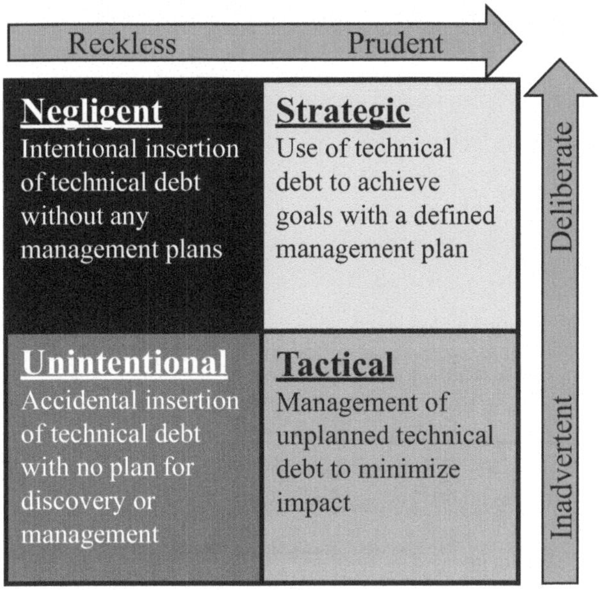

Fig. 2.2 Technical debt quadrant, adapted from [7]

on a house that you know you can afford. Unintentional debt is when the developers are reckless and unaware of what they are doing. The recklessness comes from a lack of sufficient programmatic controls: untrained developers or misunderstood requirements. However, the debt is inadvertent—it is not intentional. This debt is equivalent to a developer not understanding at the time that what they were doing was wrong. Finally, tactical debt occurs when the system meets requirements and standards, but there was a better way to have completed the work. Often only viewable in hindsight, this debt is tactical because leaving the system in this state is not necessarily a bad thing and does not necessarily incur significant interest; however, the better solution could be implemented when timing is correct (a tactical fix). Fowler's technical debt quadrant helped to shape thinking about the ways in which technical debt occurs.

We stated earlier that it is better to move up and to the right in the quadrant. This implies that negligent technical debt is better than unintentional technical debt. While both types of technical debt are reckless, negligent debt is deliberate, which means that there is some understanding of the compromise being made. This awareness of the compromise, even with no plan to repay it, is better than unintentional debt, which can remain hidden in the system. Unintentional debt can accumulate without the developer's knowledge until it is too late to avoid technical bankruptcy.

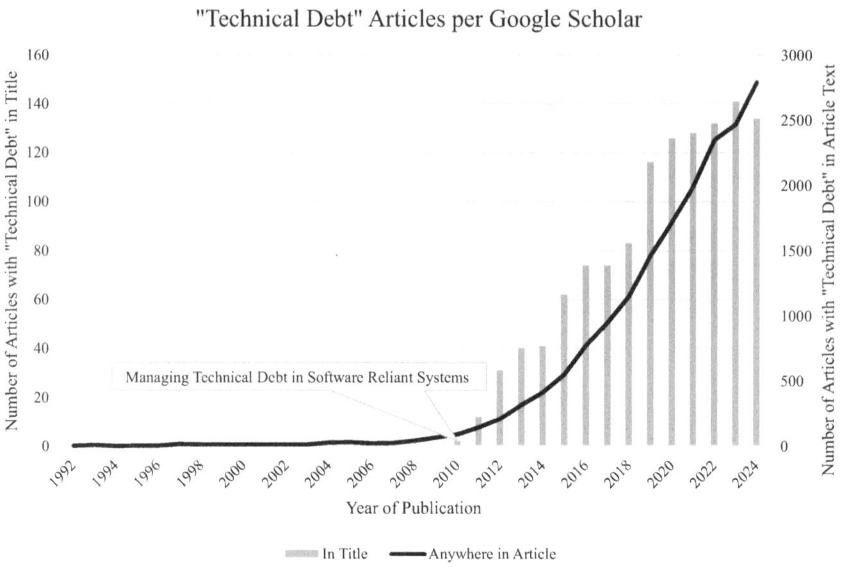

Fig. 2.3 Google Scholar search results for "technical debt" by publication year

2.1.3 Early Academic Publications

In 2010, the first workshop on managing technical debt occurred in Pittsburgh, PA [8]. This conference marked the start of substantial published discourse on technical debt, moving it from the realm of blog posts and white papers to academic literature. The paper "Managing Technical Debt in Software-Reliant Systems" by Brown et al. [9] is the first academic paper that I found that directly references technical debt in the title. Developed by authors from multiple universities and research centers, it proposes that managing technical debt needs to be part of future research in software engineering and identifies a critical research agenda for discussing technical debt moving forward.

The use of technical debt within technical articles grew exponentially following the workshop, as shown in Fig. 2.3. These data show the increasing importance of the technical debt metaphor and allow us to determine that technical debt was part of the software engineering lexicon starting in 2010.

2.1.4 Agile Development and Technical Debt

What created the surge in articles on technical debt starting in 2010? The technical debt metaphor had been around for almost 20 years at this point. Yet, here we were with an entire conference dedicated to technical debt, which was both a symptom of and a catalyst for increasing awareness of technical debt. However, the fact that technical debt was such a concern that a dedicated conference was needed implies that something else changed in the field of software development. That something was the rise of Agile software development.

In 2001, 17 software developers met in Snowbird, Utah, and created the Agile Manifesto [10], including the above-mentioned Ward Cunningham and Martin Fowler. This manifesto gave rise to Agile software development methods, which are intended to transition software development practices from heavyweight, document-driven methods with few deliveries to more responsive, lightweight methods that deliver frequently and adapt to changing requirements.

Agile methods spread rapidly and evolved beyond software development. The Project Management Institute (PMI) , started tracking the use of Agile methods in project management and their adoption by organizations as early as 2010, as shown in Fig. 2.4. The use of Agile methods steadily increased from 2010 to 2017, when hybrid methods were first tracked. Hybrid methods blend Agile methods with traditional sequential project management methods. There was a drop in reported usage from 2017 to 2020, which could be due in part to missing data from 2018 and 2019 or to the introduction of the concept of hybrid methods. The overall use of either approach in projects has steadily risen since 2020, with over 50% of organizations saying they use these methods often or always in 2023.

While not all organizations surveyed by PMI are software development organizations or systems engineering organizations, the results show a clear trend of adoption of Agile methods. Brown et al. recognized in their seminal paper [9] that

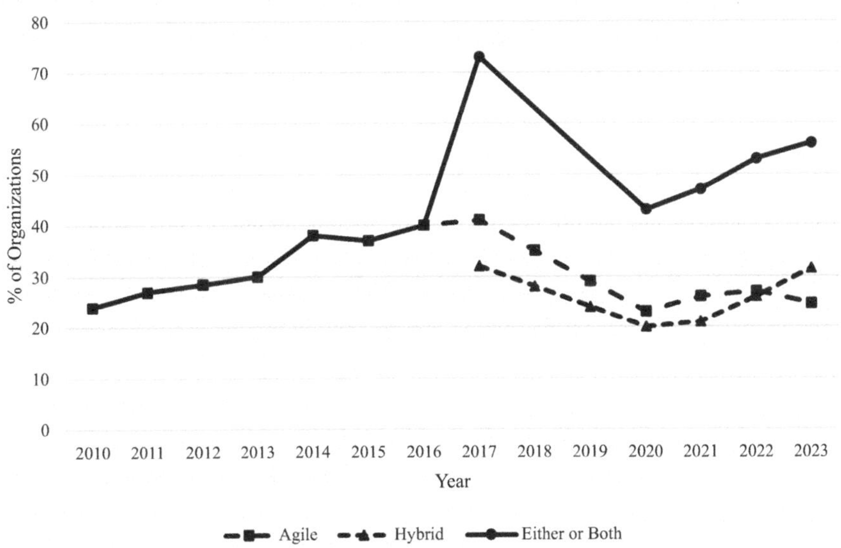

Fig. 2.4 Organizations using Agile and hybrid project management methods are often or always based on *PMI Pulse of the Profession* publications [11–23]

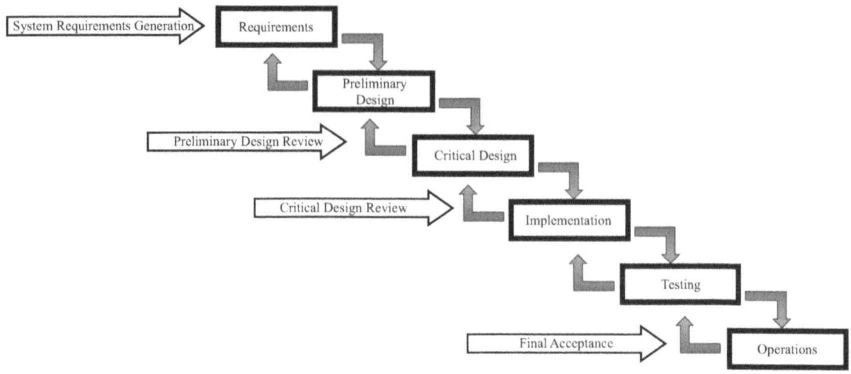

Fig. 2.5 Simplified version of Royce's sequential method, based on [24]

Agile practices had a significant impact on software development practices over "the past 5–8 years," but that technical debt management did not keep up. This statement begs the question: Is the rise in technical debt caused by the increased use of Agile methods or is it just correlated?

To determine if Agile methods have caused an increase in technical debt, we first need to understand the primary development method that was in place before the rise of Agile: sequential methods, also known as waterfall methods. First described in a 1970 paper by Winston Royce[24] and shown in Fig. 2.5, the sequential method relies on a linear progression through development steps, with each stage completing before the next one starts. The method gets its nickname from the visual depiction in Fig. 2.5, where the system lifecycle flows like a waterfall from one step to the next. Of note, modern implementations of the sequential method tend to assume a direct linear flow, while Royce identified the need for iterations, stating "Hopefully, the iterative interaction between the various phases is confined to successive steps" [24]. Royce also acknowledged that reality tends to deviate from the idealized lifecycle, with testing often resulting in redesign and adjustments to the requirements.

In this method, the requirements are set at the beginning of the design process and then remain largely unchanged throughout development. Following the requirements definition, the developer establishes a preliminary design followed by a critical design, which specifies the system design such that it can be implemented in the following stage. Following implementation, the system is tested and, if accepted, enters operations.

The transitions between phases rely heavily on documentation, and interactions with customers are limited. Royce identified three spots for stakeholder involvement in the development process: System Requirements Generation, Preliminary Design Review, and Critical Design Review. The stakeholder has final acceptance authority prior to entering Operations. These customer interaction points are indicated by the arrows in Fig. 2.5. Depending on the length of the system phases, the stakeholder may not interact with the system developer for extended periods of time, which can lead to issues with system validation. Without stakeholder inputs, the developer may build the system in accordance with its requirements (verification) but may not have built the right system for the stakeholder (validation). The linear flow through the lifecycle can also result in late discovery of changes, which can cause significant rework and cost increases [24].

Agile methods provide a different view of implementation and stakeholder involvement. The first three principles behind the Agile Manifesto [25] are:

- Our highest priority is to satisfy the customer through early and continuous delivery of valuable software.
- Welcome changing requirements, even late in development. Agile processes harness change for the customer's competitive advantage.
- Deliver working software frequently, from a couple of weeks to a couple of months, with a preference to the shorter timescale.

These principles define the fundamental differences in philosophy behind sequential methods and Agile methods: include heavy customer involvement, get frequent feedback on incremental deliveries, and be adaptable to change. The Agile Manifesto also emphasizes personal communications over design documentation. Too often, these points of emphasis are misinterpreted as Agile methods having little structure and the freedom for the developer to do whatever they want whenever they want. The opposite is actually true; the success of Agile methods is dependent upon following a highly structured process. However, the process is set up to allow for frequent changes to development tasks and adjustments to changing priorities.

We can now compare these two lifecycle models. Sequential methods are linear processes that rigidly hold to original requirements and are not adaptable to changes in requirements, resulting in a single release at the end of the project. Agile methods provide a structured process that stresses frequent releases and adaptability to change. Which of these methods is more likely to produce technical debt?

To answer this question, recall that technical debt is defined as the negative long-term impacts on the health of a system as a result of technical compromises made to achieve a short-term benefit. What would cause a developer to make a technical compromise—and do these causes differ in each lifecycle model?

In the sequential approach, the requirements, project budget, and schedule are set at the beginning of the project. Given that the requirements are not supposed to change, any issues that arise in the project must be met with either changes to schedule or budget. Assuming that delays and cost increases are acceptable, then there is no need to make a technical compromise. If the delays and cost increases are not acceptable, then the requirements of the project would need to change. These changes to requirements then flow through the documentation-heavy process, ensuring that the change is captured and the design is updated. The single delivery at the end of the project provides time to ensure that the system is correct prior to delivery. While technical debt is possible within a sequential project, its processes do not encourage the taking of shortcuts or the making of technical compromises.

Agile methods, on the other hand, encourage frequent updates to requirements based on customer interactions and frequent deliveries. The rigid delivery structure (working code at the end of every sprint) and focus on customer value can force

Table 2.1 Technical debt associated with the values of the Agile Manifesto

Agile manifesto value [26]	Technical debt types	Technical debt causes
Individuals and interactions over processes and tools	Infrastructure debt, implementation debt	Political pressure
Working software over comprehensive documentation	Architecture debt, design debt, documentation debt	Schedule pressure, cost pressure
Customer collaboration over contract negotiation	Requirements debt	Cost pressure, political pressure
Responding to change over following a plan	Domain debt, requirements debt	Cost pressure, schedule pressure, political pressure

developers to implement shortcuts to produce a high-value feature instead of spending the proper time on the less-visible supporting tasks. Changes to requirements can cause pivots in design, which, if not well documented, can lead to complicated solutions that are hard to understand and since Agile methods stress communication over documentation, documentation can often fall behind. These conditions are ripe for the development of technical debt—technical compromises that linger in the system to achieve the short-term release but have potential long-term consequences. The four values of the Agile Manifesto can be mapped directly to causes and types of technical debt, as shown in Table 2.1. We will discuss the sources and causes of technical debt in more detail in Chap. 4. We define types of technical debt in Chap. 5.

Even though Agile methods may be more likely to produce technical debt, they still provide valuable processes that can increase project success rates. The increased interaction with the customer produces better products that meet their needs, which is a good thing! The spread of Agile methods from software to program management and systems engineering shows that Agile methods have value to both the system developer and their customers. However, technical debt comes with Agile methods, and therefore we must be prepared to manage the technical debt that ensues.

2.2 Bridging from Software Engineering to Systems Engineering

In the previous section, we established how technical debt arose within the software engineering field. However, this is a book about systems engineering. Fortunately, there are many similarities between systems engineering and software engineering. Therefore, we can leverage the work done on technical debt management within software engineering as a basis for our applications to systems engineering. First, let's compare formal definitions of software engineering and systems engineering.

The Software Engineering Body of Knowledge (SWEBOK) [27] uses the ISO/IEC/IEEE definition of software engineering:

> the application of a systematic, disciplined, quantifiable approach to the development, operation, and maintenance of software; that is, the application of engineering to software [28]

Software engineering differs from computer science, in that computer science is the theoretical exploration of computing processes and algorithmic solutions, while software engineering is the application of these principles to create software.

The Systems Engineering Body of Knowledge (SEBoK) [29] uses the INCOSE definition of systems engineering:

> Systems Engineering is a transdisciplinary and integrative approach to enable the successful realization, use, and retirement of engineered systems, using systems principles and concepts, and scientific, technological, and management methods.

These definitions are similar: they both refer to approaches to produce and maintain products throughout their lifecycle and rely on information sources from other disciplines and principles. Both engineering approaches focus on the practical application of these disciplines and principles to create a final product, instead of focusing on the generation of new knowledge.

Many of the core systems engineering activities are also essential to software engineering. Both fields have lifecycles which consist of similar steps such as requirements decomposition and allocation, design, integration, verification and validation, test, and sustainment. Both fields integrate multiple components which requires detailed specification of interfaces and understanding of information flow. Model-based practices are used in both fields to drive designs. In many ways, software is a digital system. The increased dependency of physical systems on software has blurred the lines between the two disciplines [30]. Therefore, it should be no surprise that many processes and practices are shared between software engineering and systems engineering.

2.2.1 Agile Methods in Systems Engineering

The close relationship between the needs and processes of systems engineering and software engineering has allowed systems engineering to borrow from software engineering. In fact, many of the common systems engineering life cycle models originated in software engineering including the "waterfall" model [24] and the spiral model [31].

Given the commonality in approach and lifecycle models and the exchange of the lifecycle models between the field, it was only natural that agile methods would soon flow from software engineering to systems engineering. The history of acquisition guidelines from the United States Department of Defense (DoD), one of the largest single acquirers of large-scale systems, shows the adoption of different recommended systems engineering approaches in an attempt to increase on-time and on-budget delivery of working systems. Figure 2.6 shows the changes in the DoD approach, determined by reviewing the DoD 5000 series documents (both DoD

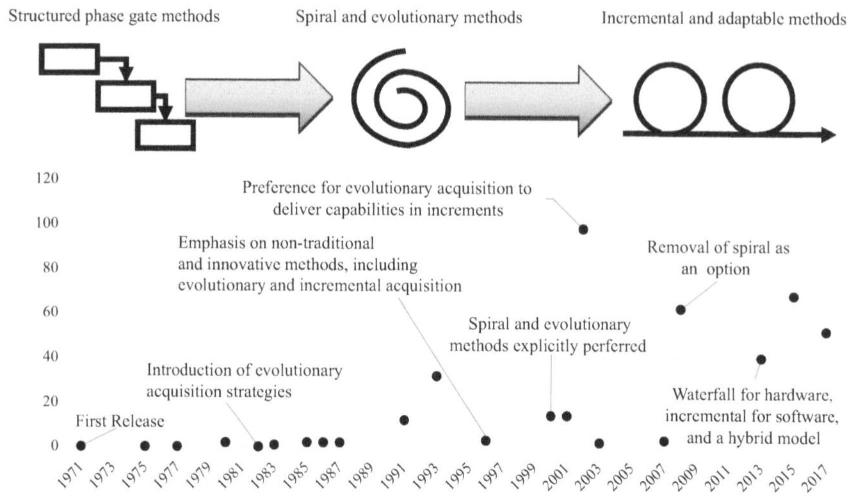

Fig. 2.6 Change of systems engineering approaches in DoD acquisition systems [7, 32]

directives (DoDD) and DoD instructions (DoDI)), which define the guidelines for the Defense Acquisition System. The points in this diagram represent the count of words related to evolutionary and incremental development, specifically "evolutionary," "incremental," and "block." In 2020, the DoD introduced the Adaptive Acquisition Framework which allowed program managers (PMs) to tailor the framework to select the best approach.

This sequence shows the arrival of Agile, incremental methods in systems engineering as a progression from sequential and spiral approaches. Published research followed this trend. In 2005, Haberfellner and de Weck presented "Agile SYSTEMS ENGINEERING versus AGILE SYSTEMS Engineering" [33] at the 15th Annual INCOSE International Symposium. In this chapter, they identified the difference between using Agile methods for systems engineering and the engineering of Agile systems, which produce flexible and adaptable systems, potentially with traditional sequential methods. The value of including agility in systems engineering processes continued to be observed, especially with the rise of Model Based Systems Engineering (MBSE). MBSE relies on a single set of data contained in the model which is dynamic in nature, instead of static documents that take time to update. MBSE enables more flexible approaches and responsiveness to change by using the model to connect all aspects of the system such that the effects of changes can be easily understood. Bruce Douglass combined Agile methods and MBSE in his book "Agile Systems Engineering" in 2015 [34].

Surprisingly, even though technical debt was a known quantity in the software engineering field by this time, and at least the association with technical debt and Agile methods was known, there is limited discussion of technical debt in systems engineering. Douglass does not discuss technical debt in his book [34]. A 2023 literature review [35] found only 18 journal publications which discussed technical

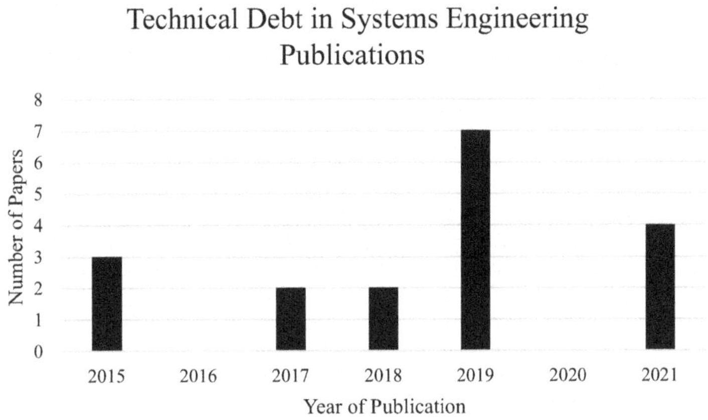

Fig. 2.7 Publication history of academic papers on technical debt in systems engineering

debt in the context of an engineering system outside of software engineering and in which technical debt in engineering was the primary focus. Figure 2.7 shows the year of publication for these sources. Of these 18 articles, 13 focused on specific systems and only 5 focused on technical debt in the broader context of systems engineering.

The International Council of Systems Engineering, INCOSE, is the primary industry group for advancing systems engineering. A March 2025 search of the INCOSE website for technical debt yielded 58 hits. The first mention in an INCOSE document is the 2015 "Project Manager's Guide to Systems Engineering Measurement for Project Success" [36], where the stated focus of the guide is to "identify and manage technical debt." Additional presentations in 2016 and 2017 mention technical debt in conjunction with a move to agile systems engineering processes. By 2023, technical debt was included in the products of the INCOSE requirements working group, reflecting a growing realization of its importance. However, the concept is still not codified in the Systems Engineering Body of Knowledge [37], as a February 2025 search returned only the following articles: "System Affordability," "Verification and Validation of Systems in Which AI is a Key Element," and "Systems Engineering and Enterprise IT."

2.3 Summary

In this chapter, we have explored the origins of the technical debt metaphor in the field of software engineering. Like many software engineering practices, such as agile development, technical debt has migrated to systems engineering. Complex systems are increasingly reliant on software and iterative development techniques and are therefore subject to increasing levels of technical debt. However, the management of technical debt is not yet a widely studied aspect of systems engineering programs.

In the next chapter, we begin to correct this oversight by developing the concepts critical for managing technical debt within systems engineering programs.

References

1. Whelan, D., & Letouzey, J. (2016, February 28). *Introduction to the Technical Debt Concept*. Agile Alliance. Retrieved April 8, 2025, from https://www.agilealliance.org/introduction-to-the-technical-debt-concept/
2. Cunningham, W. (1992, March 26). *The WyCash Portfolio Management System*. Retrieved January 29, 2022, from http://c2.com/doc/oopsla92.html
3. Investopedia. (n.d.). *Dictionary*. Investopedia. Retrieved April 8, 2025, from https://www.investopedia.com/financial-term-dictionary-4769738
4. Gorton, D. (2024, October 31). *Taxes Definition: Types, Who Pays, and Why*. Investopedia. Retrieved June 24, 2025, from https://www.investopedia.com/terms/t/taxes.asp#:~:text=Taxes%20are%20mandatory%20contributions%20levied,as%20Social%20Security%20and%20Medicare
5. McConnell, S. (2008, June). *Managing Technical Debt*. Construx Software Builders. Retrieved January 27, 2022 from https://www.construx.com/uploadedfiles/resources/whitepapers/Managing%20Technical%20Debt.pdf
6. Fowler, M. (2009, October 14). *TechnicalDebtQuadrant*. Retrieved January 27, 2022, from https://martinfowler.com/bliki/TechnicalDebtQuadrant.html
7. Kleinwaks, H. (2023). *Avoiding Technical Bankruptcy in System Development: A Process to Reduce the Risk of Accumulating Technical Debt*. Doctoral Dissertation, Colorado State University, Department of Systems Engineering.
8. Izurieta, C., Ozkaya, I., Seaman, C., Kruchten, P., Nord, R., Snipes, W., & Avgeriou, P. (2016). Perspectives on Managing Technical Debt: A transition point and roadmap from Dagstuhl. *Joint of the 4th International Workshop on Quantitative Approaches to Software Quality, QuASoQ 2016 and 1st International Workshop on Technical Debt Analytics, TDA 2016*, 84–87.
9. Brown, N., Cai, Y., Guo, Y., Kazman, R., Kim, M., Kruchten, P.,... & Zazworka, N. (2010). Managing Technical Debt in Software-Reliant Systems. *Proceedings of the FSE/SDP workshop on Future of software engineering research*, 47–52.
10. Highsmith, J. (2001). *History: The Agile Manifesto*. Agile Alliance. Retrieved April 8, 2025, from https://agilemanifesto.org/history.html
11. Project Management Institute. (2011). *PMI's Pulse of the Profession: Highlighting key trends in the project management profession*. (Pulse of the Profession:2011).
12. Project Management Institute. (2012). *PMI's Pulse of the Profession: Driving Success In Challenging Times*. (Pulse of the Profession:2012).
13. Project Management Institute. (2013). *PMI's Pulse of the Profession: The High Cost of Low Performance*. (Pulse of the Profession:2013).
14. Project Management Institute. (2014). *PMI's Pulse of the Profession: The High Cost of Low Performance*. (Pulse of the Profession:2014).
15. Project Management Institute. (2015). *PMI's Pulse of the Profession: Capturing the Value of Project Management*. (Pulse of the Profession:2015).
16. Project Management Institute. (2016). *PMI's Pulse of the Profession: The High Cost of Low Performance*. (Pulse of the Profession:2016).
17. Project Management Institute. (2017). *PMI's Pulse of the Profession: Success Rates Rise Transforming the High Cost of Low Performance*. (Pulse of the Profession:2017).
18. Project Management Institute. (2018). *PMI's Pulse of the Profession: Success in Disruptive Times*. (Pulse of the Profession:2018).
19. Project Management Institute. (2019). *PMI's Pulse of the Profession: The Future of Work*. (Pulse of the Profession:2019).

20. Project Management Institute. (2020). *Ahead of the Curve: Forging a Future-Focused Culture.* (Pulse of the Profession:2020).
21. Project Management Institute. (2021). *Beyond Agility.* (Pulse of the Profession:2021).
22. Project Management Institute. (2023). *Pulse of the Profession 2023: Power Skills, Redefining Project Success.* (Pulse of the Profession:2023).
23. Project Management Institute. (2024). *Pulse of the Profession 2024: The Future of Project Work: Moving Past Office-Centric Models.* (Pulse of the Profession:2024).
24. Royce, W. (1987). Managing the development of large software systems: concepts and techniques. *Proceedings of the 9th international conference on Software Engineering,* 328–338.
25. Agile Alliance. (n.d.). *Principles behind the Agile Manifesto.* Agile Alliance. Retrieved April 8, 2025, from https://agilemanifesto.org/principles.html
26. Agile Alliance. (n.d.). *Manifesto for Agile Software Development.* Agile Alliance. Retrieved April 16, 2026, from https://agilemanifesto.org
27. Washizaki, H. (Ed.). (2024). *Guide to the Software Engineering Body of Knowledge (SWEBOK Guide), 4.0.* IEEE Computer Society. Retrieved March 7, 2025 from https://www.swebok.org
28. International Organization for Standardization/International Electrotechnical Commission/ Institute of Electronic and Electrical Engineers. (2017). *Systems and Software Engineering - Vocabulary, 2nd Edition.* (ISO/IEC IEEE Standard 24756:2017). https://www.iso.org/standard/ 71952.html
29. SEBoK Authors. (2024, November 25*). Systems Engineering Overview.* SEBoK. Retrieved April 18, 2025, from https://sebokwiki.org/w/index.php?title=Systems_Engineering_Over view&oldid=72764
30. SEBoK Authors. (2024, November 25). *Systems Engineering and Software Engineering.* SEBoK. Retrieved May 21, 2025, from https://sebokwiki.org/w/index.php?title=Systems_E ngineering_and_Software_Engineering&oldid=72539
31. Boehm, B. W. (1988). A Spiral Model of Software Development and Enhancement. *Computer,* 61–72. https://doi.org/10.1109/2.59
32. Mortlock, R. F. (2020). Studying acquisition strategy formulation of incremental development approaches. *Defense AR Journal*, 27(3), 264–311.
33. Haberfellner, R., & de Weck, O. (2005). Agile SYSTEMS ENGINEERING versus AGILE SYSTEMS engineering. *Fifteenth Annual International Symposium of the International Council On Systems Engineering (INCOSE).*
34. Douglass, B. P. (2016). *Agile Systems Engineering.* Morgan Kaufmann.
35. Kleinwaks, H., Batchelor, A., & Bradley, T. H. (2023). Technical Debt in Systems Engineering - A Systematic Literature Review. *Systems Engineering, 26*(5), 675–687. https://doi.org/10. 1002/sys.21681
36. Carson, R. S., Frenz, P. J., & O'Donnell, E. (2015). *Project Manager's Guide to Systems Engineering Measurement for Project Success: A Basic Introduction to Systems Engineering Measures for Use by Project Managers (Version 1.0).* INCOSE.
37. SEBok Authors. (2024, November 25). *Guide to the Systems Engineering Body of Knowledge (SEBoK).* SEBoK. Retrieved February 5, 2025, from https://sebokwiki.org/w/index.php?title= Guide_to_the_Systems_Engineering_Body_of_Knowledge_(SEBoK)&oldid=73360

Chapter 3
Key Concepts for Managing Technical Debt in Systems Engineering

Abstract The differences between systems engineering and software engineering are sufficient enough to require a systems engineering centered definition of technical debt. Additionally, a consistent ontology for technical debt within systems engineering is required. This ontology is provided along with a visual explanation of technical debt through a context map. Several items that are often presented as technical debt but which do not meet the ontological definitions are provided. The effect of technical debt on the three system dimensions of performance, cost, and schedule, is discussed, showing how achieving a short-term benefit in one dimension requires making compromises in another dimension. Finally, the concept of technical credit is introduced.

Keywords Technical debt definition · Technical debt ontology · Technical debt concept map · Principal · Taxes · Interest · Fees · System dimensions · Technical credit

So far, we've seen that while technical debt does occur within systems engineering projects, it is not widely cited in academic papers or used as part of the everyday lexicon of practicing systems engineers. The lack of discussion may be due to systems engineers not having a common understanding of how to talk about technical debt. What terms should be used to talk about the effects that we call technical debt? How does technical debt impact the development of a system?

In this chapter, we start to provide answers to these questions by introducing an ontology for technical debt within systems engineering. We then develop a technical debt context map, showing the conceptual relationships between the items in the ontology. Finally, we discuss systems engineering projects from the viewpoint of three system dimensions (cost, schedule, and performance) and show how this viewpoint leads to an understanding of the impacts of technical debt.

H. Kleinwaks, *Managing Technical Debt in Systems Engineering*,
https://doi.org/10.1007/978-3-032-30360-8_3

3.1 Refining the Definition of Technical Debt for Systems Engineering

Before reviewing the full ontology for technical debt in systems engineering, we first need to revisit the definitions that were provided in Sect. 1.1. Here, we introduced both a metaphorical and a quantitative definition of technical debt. When used as a metaphor, technical debt serves to illuminate the potential consequences of decisions. This use is helpful in broader discussions about possible compromises within the development of a system. However, technical debt is also a measurable quantity used to evaluate the health of the system, whether measured in dollars, increased work hours, or increased system risk.

The following concepts hold true for both the metaphorical and quantitative definitions of technical debt:

1. Technical debt is the result of a decision made for short-term benefit with the potential for long-term negative impacts.
2. Technical debt is the result of a technical compromise: sacrificing budget or schedule to achieve a benefit is not technical debt.
3. Technical debt arises due to a known event: the technical compromise has already occurred.
4. The impact of technical debt is uncertain—the long-term impact may or may not be realized.
5. Technical debt may be beneficial to a project, but excess technical debt can bring a project to a complete standstill.

We will discuss each of these concepts in the following paragraphs.

Technical debt is accrued for a short-term benefit

Throughout system development, the developer is faced with decisions. Many of these decisions produce an immediate benefit, such as enabling the next step in the project to proceed. Other decisions may have long-term benefits, such as investing in increased infrastructure to support future manufacturing needs. So why are only decisions made for short-term benefits considered to be technical debt? It comes down to a question of planning. Any decision made to produce a long-term benefit occurs as the result of planning. To recognize long-term benefits, there must be a strategic plan in place to achieve that benefit. Decisions made for short-term benefits, due to compressed time frames, often have reduced planning and analysis associated with them, and therefore are more likely to produce negative long-term consequences. This aspect is the other key part of this technical debt concept: technical debt results from decisions with the potential for negative long-term consequences. These consequences may never be realized, and the impact of the consequences may vary. However, it is critical to track the potential for long-term consequences to prevent them from accumulating within the system. To count as technical debt, the consequences must be long-term such that there is the potential for delayed impacts.

A short-term benefit with short-term consequences is not technical debt because it is quickly resolved.

Technical debt is the Result of a Technical Compromise

Technical debt must be the result of a technical compromise, which is a compromise that reduces the performance of the system. If a decision is made to borrow more money to add staff to meet a deadline, then the profitability of the system may be reduced. However, the end product is the same; it just costs more money to get there. If a decision is made to delay a release to ensure full performance, then the system is late, but still meets performance requirements. While these cases could result in financial debt, they can be handled through well-known project management methods. Technical debt, however, introduces variable performance consequences and does not have established management methods. We will restrict our discussion of technical debt to the results of technical compromises.

A technical compromise is not always the change in the system design or implementation, although that is often easier to envision. The technical compromise may occur instead when the change to the design is rushed into the system without proper consideration of its impacts on the rest of the system performance. For example, the decision to use a cheaper part (short-term benefit is reduced cost) does not necessarily incur technical debt if an analysis shows that performance is not impacted. However, technical debt would be incurred if the performance impacts occur and are not mitigated through additional changes to the system.

The technical compromise has occurred

Unlike risk management, where the triggering event may or may not occur, technical debt is the result of a compromise which has occurred. Using the technical debt quadrant (Fig. 2.2), this compromise may be deliberate (intentional) or inadvertent (unintentional). In either case, the compromise was made and the system developer must handle the consequences. Proper technical debt management requires that technical debt is identified when a deliberate compromise is made and that techniques are regularly applied to search for inadvertent technical debt in the system and to mitigate its impact.

The impact of technical debt is uncertain

One of the toughest things about technical debt is that its impact is uncertain. There may be part of the system with significant technical debt: poor documentation, incomplete interface identifications, reduced usability, etc. However, if that part of the system does not need to be upgraded or changed, then the potential impacts on the system may never be realized. Therefore, the "interest" on the debt is a probabilistic quantity, and the likelihood of its occurrence must be accounted for when managing the technical debt.

Technical debt may be beneficial, but too much can cripple a system

Not all technical debt is bad. With proper management techniques, the system and the developer may realize substantial benefits from taking on technical debt to meet a critical milestone, such as being first to market with a new product. However, if left unchecked, technical debt can cause technical bankruptcy: the state where the project cannot proceed without paying down some or all of its technical debt, potentially resulting in project cancellation.

3.2 Ontology of Technical Debt

Before diving into the concepts underpinning technical debt management in systems engineering, we need to establish an ontology to define the key terms and establish their relationships [1]. Table 3.1 provides the definitions of the key terms, and in the next section we will provide a technical debt concept map to establish the relationships.

3.3 Technical Debt Concept Map

A context map of technical debt in systems engineering is shown in Fig. 3.1. The context map is a high-level relationship diagram that defines interactions between the stakeholders and system components. By walking through this map, we can understand some of the ways in which technical debt can be created and how it can impact the system development. Understanding the flow of technical debt throughout the system will enable us to develop management strategies in the rest of this book.

With infinite time and money, there would be no reason for a system developer to make a technical compromise. However, both time and money are finite, and systems are developed in a business context. Therefore, system developers are beholden to the business goals of the stakeholders. The concept map starts with these goals in the upper left. The stakeholders want to achieve their business goals, which induces them to exert pressure on the system developers. To relieve that pressure, the system developers are forced to make a technical compromise which yields a short-term benefit to satisfy the stakeholders and allow them to achieve their business goals. This path is shown as Loop 1 (in dark gray) in Fig. 3.1.

If there were no concerns about the impact of the technical compromise, the context map would stop with just Loop 1. However, the technical compromise may create technical debt, which is composed of technical debt items. Each technical debt item causes a long-term consequence, which negatively impacts the health of the system. The reduction in system health negatively impacts the stakeholder's ability to satisfy their business goals. This loop is shown as Loop 2 (in light gray) in Fig. 3.1.

Loop 2 is a negative feedback loop—as the business goals are negatively impacted, the stakeholders may exert more pressure on the system developers, causing them to make additional technical compromises and further reducing the health of the system.

Table 3.1 Technical debt ontology, adapted from [2]

Term	Definition	Discussion
Artifact	Part of the system which is affected by the technical debt	Associating technical debt with a specific artifact (such as an interface, a component, or the entire system) enables the developer to understand which parts of the system are impacted
Balance	Cost to repay the system technical debt	The balance, B, is a measurement of how much it costs to repay the system technical debt at the present time
Benefit	Positive outcome of a compromise	A benefit increases the distance between a system dimension vertex and the center of the system dimension triangle or prevents the reduction of that distance, thereby increasing or preserving the system value
Cause	Reason why a technical compromise was made	The cause identifies the source of the pressure to make the compromise and the short-term benefit
Cause category	System dimension where short-term benefits are achieved	The cause category allows identification and traceability of similar technical debt items
Consequence	Result of the technical compromise	The consequence, often negative, describes the predicted impacts of making the technical compromise, in any of the system dimensions
Description	Narrative of the technical debt item	The description provides amplifying information about the technical debt item, including the rationale for the technical compromise, the achieved benefits, and the potential consequences
Fee	Reduction in the performance dimension experienced by the user of the system due to technical debt	Fees represent the performance cost to the user, such as delayed service time or lack of fidelity in results, that occurs with each use of the system due to technical debt. A more detailed discussion of fees is in Sect. 3.4
Fee amount	Amount of performance reduction with each use of the system	The fee amount, g, measures how much performance is lost with each fee instance
Fee frequency	The rate at which the user pays the fee	The fee frequency, f, measures how often the fee is paid by the user
Interest	Expected cost to work around unpaid principal	The interest, I, is a measure of the extra work required to continue the development due to technical debt. A more detailed discussion of interest is in Sect. 3.4

(continued)

Table 3.1 (continued)

Term	Definition	Discussion
Interest amount	Amount of additional work associated with a technical debt item	The interest amount, a, is a function of the state of the system and represents the amount of work to complete system development in the presence of a technical debt item
Interest probability	The likelihood that the interest amount will be realized	The interest probability, r, represents the probability that the system developer will need to pay the interest amount for a particular technical debt item
Long-term	A lasting effect	Long-term benefits and impacts may take extended periods of time to be realized or may last for extended durations
Negative impact	A reduction of the system health in at least one of the system dimensions	Negative impacts are typically associated with technical compromises and may occur in any of the system dimensions
Principal	A measurement of the compromise in the performance dimension	The principal, P, represents the amount "borrowed" from the future system to achieve the short-term benefit. Measuring the principal is discussed in Chap. 9
Specific cause	Short-term benefit produced by the technical compromise	The specific cause is the actual benefit achieved by making the technical compromise
Short-term	An immediate (or near immediate) time frame	Short-term benefits and impacts occur rapidly after making the compromise
System dimensions	Three interrelated aspects of system development: cost, schedule, and performance	The system dimensions provide a method to discuss the interrelationships between different aspects of system development and the associated impacts of technical debt
System health	Ability of the system to accomplish its objectives in all three system dimensions	A healthy system can complete its performance objectives on time and on budget. Impacts to the system health may occur in any of the dimensions—increasing cost, delaying delivery, or reducing performance
System technical debt	Aggregation of all technical debt items	System technical debt is used to distinguish the actual instances of technical debt within a system from the metaphorical definitions
Taxes	Additional work or cost required when principal is repaid	Repaying the principal often requires the undoing of temporary fixes or work-arounds. Taxes, T, represent this extra effort required to undo the work to enable the implementation of the correct solution. A more detailed discussion of taxes is in Sect. 3.4

(continued)

Table 3.1 (continued)

Term	Definition	Discussion
Technical bankruptcy	State where the system can no longer proceed with its lifecycle until technical debt is repaid	Technical bankruptcy, discussed in Chap. 12, occurs when the consequences from technical debt are so severe that the system cannot progress in its lifecycle without pausing to repay some, or all, of the technical debt items
Technical compromise	Concession made in the performance dimension	Technical compromises are made in either scope or quality of the system and represent the foundation of technical debt items
Technical credit	The inverse of technical debt	Technical credit is the inverse of technical debt—a benefit in the performance dimension on the long-term health of the system due to a compromise made in any of the system dimensions
Technical debt (as a measurable system parameter)	The measurable impact on the long-term health of the system due to technical compromises made for short-term benefit	Quantification of the technical debt as impacts to the system health allows for estimation of the time and cost to repair the system and measurement of the consequences of inaction
Technical debt (as a metaphor)	A reflection of the idea that compromises made for short-term benefit can have long-term impacts on the system health	The metaphorical definition provides context for discussions and is easily relatable to professionals from all fields
Technical debt item	A specific instance of technical debt in the system	The technical debt item is the entity used to record and manage instances of technical debt in the system and is scoped to a single technical compromise and its associated impacts
Total cost	Summation of balance and interest across all technical debt items	The total cost, C, represents the aggregated cost to the developer to repay all technical debt items
Total fees	Summation of fees across all technical debt items	The total fees, F, represent the aggregated performance cost to the user
Type	Classification based on the affected artifacts	Types are defined based on artifacts (see Chap. 4), and grouping items by type allows for development of specific management strategies

(continued)

Table 3.1 (continued)

Term	Definition	Discussion
Unit	Measurement of change in the performance dimension of the system	Consistent units are required across different technical debt items to ensure that they can be compared
Value	Measurement of the value of the system to the stakeholders	Represented as the area of the system dimension triangle, the value of the system is an expression of the satisfaction of the stakeholders with respect to the system delivery

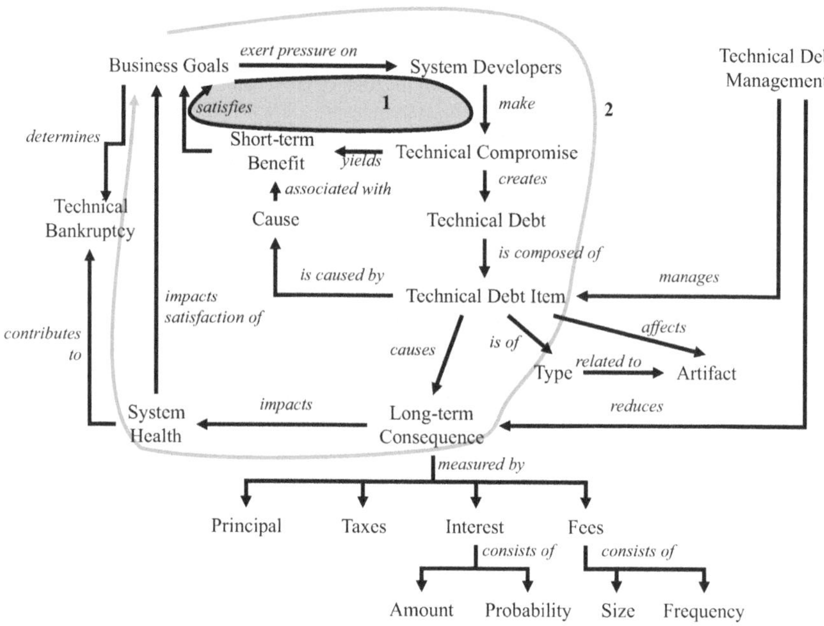

Fig. 3.1 Context map of technical debt for systems engineering adapted from [2]

Eventually, the system health is in a state which pushes the system into technical bankruptcy.

Using the context map, we see that the long-term consequences negatively impact the system health which negatively impacts the stakeholder's business goals. Since the ability to satisfy their goals drives stakeholders to put pressure on the developer, it is critical to manage the entire technical debt lifecycle.

3.4 Principal, Taxes, Interest, and Fees

The bottom of the concept map (see Fig. 3.2) shows how the long-term consequences of the technical compromises can be measured through four different quantities: principal, interest, taxes, and fees. These terms, in keeping with the financial analogy, are used to represent how much has been borrowed from the future health of the system and how much will need to be repaid. Let's start by looking at principal and taxes, which are two components of technical debt that must be repaid by the system developer to remove the technical debt.

The principal represents the amount "borrowed" from the future health of the system and is measured as the size of the concession in the performance dimension. This means that principal can be thought of as the amount of time or money saved by not implementing the "correct" solution. For example, consider a decision regarding updating an operating system (OS) for a cyber-physical system prior to its release. It will take time to ensure that all of the services work with the new version of the OS, which will delay the release. However, using the old OS may leave the system vulnerable to cybersecurity threats. Making the decision to use the old OS to meet the release date is a technical compromise, and the principal can be measured as the time it would have taken to update all of the services to deploy with the new OS.

The implementation of a technical compromise often involves the use of non-ideal solutions. Therefore, repaying the technical debt requires not only implementing the "correct" solution (repaying the principal) but also removing the non-ideal solution.

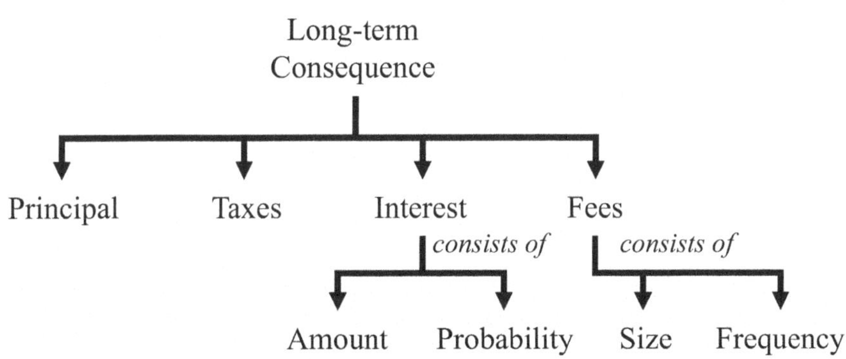

Fig. 3.2 Measurement of long-term consequences

The removal of the non-ideal solution is a tax[1] on the system developer: extra work on top of the work already required to be performed.

Principal and taxes are paid by the system developer through increased development timelines and effort and can be estimated at the time the technical debt is identified. The impact of the technical compromise on the development of other parts of the system and its utility to users is less easily determined and must account for the likelihood that problems occur.

Interest defines the extra work required to update or continue development in the presence of existing unpaid principal [4–8]. These impacts affect the system developer—the developer must work harder to add new capability because of the unpaid principal. However, it is not certain that interest will occur. If we consider the same case defined above with the release of the old OS, the interest would be any work required to update new applications to work with the old OS or to implement additional security patches for the old OS. If the new applications are backward compatible or no new cyber threat is discovered, then no new work is required and no interest would need to be paid. On the contrary, if a new exploit is revealed specific to the old OS, then there would be a significant amount of interest required to patch and update the old OS. Based on the non-deterministic nature of interest, it is broken down into an amount (the amount of work to do) and a probability.

Principal, taxes, and interest will all be paid by the system developer—the system developer needs to spend additional time and effort in the continued development and maintenance of the system. However, technical debt can also affect the user through the delivery of a system that is less performant, less maintainable [9, 10], or less usable. For example, the use of an old OS may cause slower runtimes and more frequent errors. Every time the user uses the system, they will have to confront these issues, which are defined as the technical fees. Each use of the system results in underperformance for the user. Therefore, the fees can be measured through a fee amount and a fee frequency—how bad is the underperformance and how often does it occur. A technical debt item with a large fee amount but a low fee frequency could be more acceptable to the user than one with a lower amount but larger frequency.

The system lifecycle stages where interest, taxes, and fees are paid are shown in Fig. 3.3. Interest affects the cost and time to update and deploy new versions of the system and is paid by the developer in all stages of the lifecycle as the developer may always have to work around technical debt. Fees affect the use of the system and are paid by the user in operations, once the system is deployed. Taxes are paid throughout the lifecycle whenever the principal is repaid and additional work is required to remove the workarounds put in place due to the technical debt. This makes taxes cyclic—the repayment can occur anywhere in the system development lifecycle and

[1] Ryan Noguchi introduced the concept of taxes [3] in conjunction with technical debt. He defined taxes as the work required to work around the technical compromise and the interest as the work required to correct the implementation. However, I feel that this definition is not consistent with previous work in the field and that the probabilistic nature of interest better suits it to reference the cost of implementing the workaround (which is unknown). Using taxes to represent the work to undo the workaround allows them to be a fixed quantity, once the workaround has been applied, and therefore they are not probabilistic.

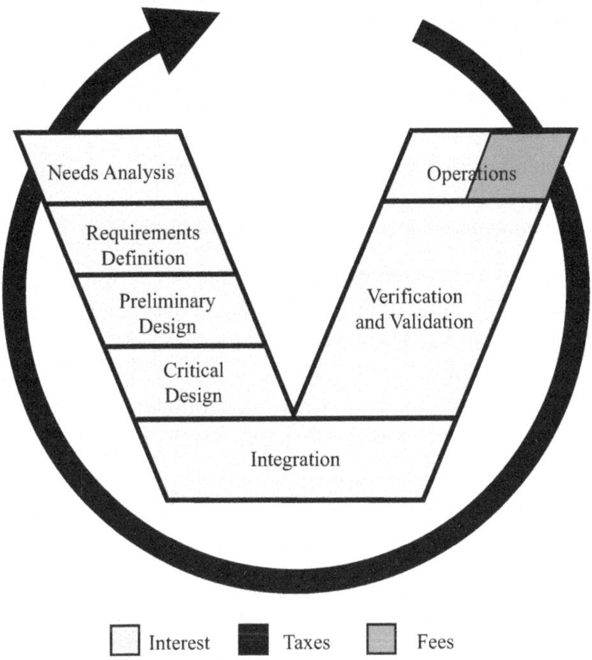

Fig. 3.3 Stages in system lifecycle where interest, taxes, and fees are paid

may require moving back through multiple stages. We will cover techniques for quantifying principal, interest, taxes, and fees in Chap. 9.

3.5 What is not Technical Debt

We've spent most of this chapter discussing the definition of technical debt as it applies to systems engineering. However, not everything that is wrong in a system is because of technical debt. If we over-generalize such that everything is technical debt, then we cannot identify distinct management strategies to help solve the problems. On the other hand, if we limit our definition such that almost nothing is technical debt, then we are left with a management strategy that only applies in very specific situations. We need to ensure that our definition is useful to system developers to allow them to develop effective management strategies. This means not over-generalizing or over-specifying technical debt [11].

Therefore, this section takes a look at how to distinguish technical debt from situations that may seem like technical debt, but should be categorized differently. First, let's start with the realization that not every technical problem represents technical debt. Just because there are bugs in the code or errors in the design does not mean that there is technical debt; sometimes this is just poor work. The definition of technical debt in systems engineering leads to the following criteria for determining if an issue is technical debt:

- A technical compromise (intentional or unintentional) was made to achieve a short-term benefit.
- There is the potential for a negative long-term consequence to the system or its users.
- The technical compromise affects one or more of the system artifacts.

These criteria allow us to examine specific situations of interest, some of which have been categorized by other authors as technical debt, and to determine if they are, indeed, technical debt. The INCOSE PM Guide to SE Management [12] defines technical debt as "the promise to complete a technical shortcoming in the future while declaring it complete today" and cites the following example involving the submission of an item on the contract data requirements list (CDRL):

> The schedule has as a task "deliver a CDRL," typically information in some form. The CDRL item is delivered, so the Control Account Manager (CAM) takes 100% credit and moves to the next task. The schedule was not developed to have a separate task for CDRL approval with duration and budget. Depending on your customer, that CDRL may require rework for any of a number of reasons, such as missing or incorrect content. That unplanned rework is technical debt.

Let's examine this example through the criteria identified above:

- Technical compromise: incorrect content in the CDRL item
- Short-term benefit: delivering the CDRL item on time
- Long-term consequence: there is rework required to correct the CDRL item, but presuming that the rework is successful, there would be no other long-term impacts
- System artifacts: the CDRL item and any components which depend upon it.

As written up in the guide, this example does not constitute technical debt, per our definition. The schedule is delayed due to the need to perform rework on the CDRL item, but the project manager should have factored in a review and rework cycle into the schedule. There are no long-term consequences since the errors in the CDRL item are rapidly corrected and do not affect the ability to complete future tasks. This situation actually represents a good management and review process and prevents technical debt from entering the system. The result is unplanned rework and a hit to the schedule, but that unplanned rework is not technical debt.

If the CDRL item rework is not performed, then we have technical debt. In this case, the technical compromise is skipping the rework, which puts errors into the system. These errors may complicate future designs and interfaces. These complications are the interest on the technical debt. Correcting the errors then involved fixing the CDRL item and removing any workarounds in the new components, which causes even more total work.

Table 3.2 provides additional examples of issues that are often confused with technical debt. The categorization of an issue as technical debt or not is never completely clean, and the "Could be technical debt if" column attempts to provide examples of when this issue could become technical debt. By insisting that issues meet all the criteria above to be considered technical debt, we ensure that technical debt management strategies can be identified across those criteria, and therefore those strategies would apply to all instances of technical debt.

Table 3.2 Examples of items that are not technical debt. Adapted from [13]

Issue	Definition	Why it is not technical debt	Could be technical debt if
Incomplete work	Planned but unfinished work [14], incomplete planned or unplanned new features or functions [11]	No technical compromise, no short-term benefit, no long-term consequences; work is planned but incomplete or not yet on the schedule	Technical compromise made to release the system with incomplete work and this incomplete work affects the development of other components
Defects and failures	System errors, such as software bugs, and poor work resulting in defect reports [15]	No short-term benefit achieved; negative impact on the system is immediate instead of long-term [16]	Compromises made to limit testing or defect resolution to save schedule; misinterpretation of requirements lead to defects
Technical compromise with no long-term impacts	Technical trade-offs reduce performance in one area in favor of another, but all system requirements met	No negative long-term impacts since the system meets all requirements and objectives	Unexpected impacts of the technical trade off result in failure to meet requirements
Poorly designed or implemented systems	Systems that pass verification (built the system right) but fail validation (built the wrong system)	No short-term benefit achieved; no technical compromises made	Compromises made in understanding of the customer's needs led to the incorrect or incomplete requirements specifications
Schedule and budget omissions	Failure to account for all tasks in project budgets and schedules [12]	Long term consequences increase the schedule/budget to account for missed tasks, but do not negatively impact the overall system health	Unaccounted for tasks are never completed, resulting in significant impacts on future development activities
System risks	Potential inability to achieve project objectives [17]	A risk is something that may happen in the system and is not the result of a technical compromise and does not produce a short-term benefit	Technical debt occurs when risks are accepted, see Chap. 7

3.6 System Dimensions

In project management, there is the concept of the "Iron Triangle" [18]. This triangle represents the interconnectedness between a project's cost, schedule, and scope/quality. The saying is "You can have it good, fast, or cheap—pick two." This idea can be migrated to systems engineering by characterizing systems engineering projects along three similar dimensions: budget, schedule, and performance [2]. Similarly to the "Iron Triangle," the three dimensions are interconnected and must be kept in balance with each other [19], as shown in Fig. 3.4. A stakeholder can define two of the three dimensions and the system developer then determines the third dimension such that the overall system provides them with sufficient value. For example, the stakeholder can define the scope through a statement of work and specify the delivery timeline. The system developer proposes a cost to the stakeholder such that they can finish the project on time and still gain value, which may be profit. The system value is not limited to profit and may be determined by a combination of factors across the three dimensions. Results such as being first to market or developing unique technology provide value to both the system developer and stakeholders.

The triangle shown in Fig. 3.4 can be used to explore this relationship. Each vertex represents one of the system dimensions: performance (P), budget ($), and schedule (T). The area of the triangle represents the value of the system to its stakeholders. Expanding any one of the vertices away from the center of the triangle will increase the value of the system by delivering it to market faster (T), more cheaply ($), or with increased performance (P). The farther each vertex is from the center of the triangle, the more positive the impact is to the stakeholder and the larger the derived value.

The performance dimension includes both scope and quality. Scope is the amount of work included in the project, typically specified through the system requirements. An increase in scope, if achieved, will increase the overall performance of the system.

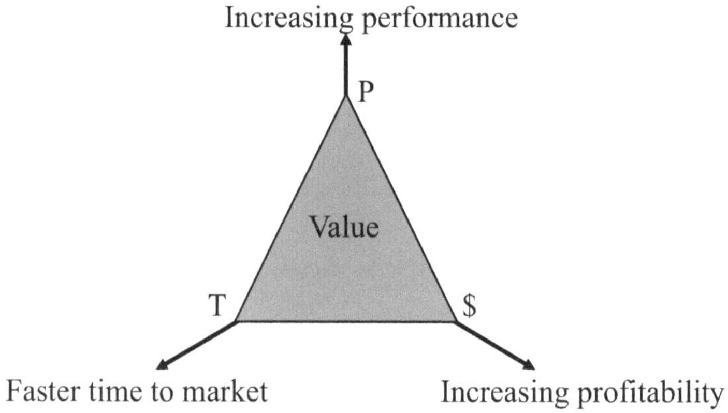

Fig. 3.4 Interconnected system dimensions showing an estimation of system value, adapted from [2]

However, this increase in scope often comes at the expense of additional time or money. Quality is a measure of how well the system meets the user's needs [20] and measures how well the system scope is achieved. High-quality projects will include multiple controls and tests to ensure that the system meets the requirements in various corner cases and off-nominal operating conditions. These projects will have highly repeatable processes and techniques. Like additional scope, adding quality costs both time and money.

The triangle visualization in Fig. 3.4 is useful when trying to understand the impact of changes in value and if a change is worth the required cost. Let's work through a simple example. Imagine that you are developing a new electric car. The design meets the initial requirements for range and the associated system development triangle is shown in Fig. 3.5a. However, the stakeholders are informed that the competitor's design will have a range that is 30 miles more than your car and they request that your design be modified to increase the range by 50 miles. This request pulls the performance vertex of the system dimension triangle up, increasing the overall value of the system, as shown in Fig. 3.5b.

However, this is not a free trade—to increase the range of the electric car, you need to increase the size of the batteries, offset the increased battery weight by using more carbon fiber in the structure, and redesign the body shape of the car to improve the aerodynamics and reduce drag. These changes take time to execute, which pushes the schedule vertex towards the center of the triangle, as shown in Fig. 3.5c. Additionally, these design changes increase the cost to manufacture the car. The price point cannot increase or else the car will be too expensive for its market. Therefore, the profitability of the car decreases, and the budget vertex moves closer to the center of the system dimension triangle, shown in Fig. 3.5d. The triangle now has gotten significantly narrower, showing an overall decrease in the value of the system.

3.6.1 Technical Debt in the System Dimensions

To see how technical debt affects the system, consider the following example. You are developing a system using an adaptive/agile system lifecycle. It is coming up on the end of the sprint and the system will be delivered to the user after the sprint completion. You are responsible for executing a test during this sprint to show that the system is ready for release. There is a lot of pressure from the system stakeholders to release the system to meet promised marketing deadlines. However, the test procedures are not complete and all tests cannot be executed. This baseline state is shown in Fig. 3.6a.

What are the options here? You can wait until the test procedures are finished and all tests are executed prior to releasing the system. This choice maintains system performance, but it has a negative impact on the system schedule. There is an associated impact on the profitability of the system: The late release results in additional development costs and can also impact future sales. Therefore, the performance vertex in the systems dimension triangle stays in place, but the schedule and cost

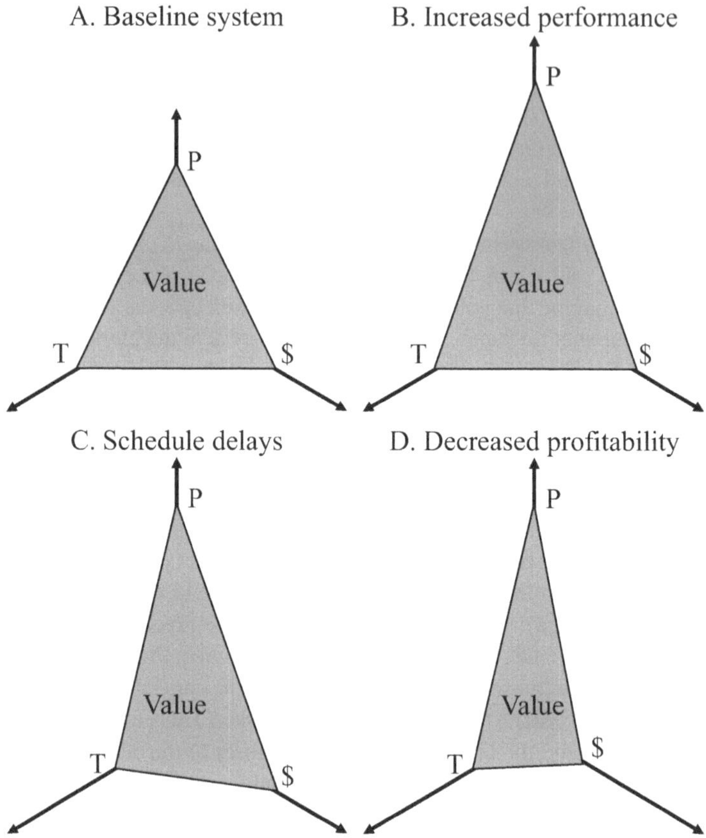

Fig. 3.5 Effect of changes in the system dimensions on the value of the system

vertices move toward the center of the triangle, decreasing the overall value of the system, as shown in Fig. 3.6b. The dashed lines indicate the starting position of a vertex along a dimension, while the solid lines indicate the ending position of that vertex.

Another choice is to perform the tests that can be performed and release the system as-is. This decision achieves the short-term benefit of maintaining the schedule due to the technical compromise on the level of testing. Therefore, the schedule vertex stays put and the performance vertex contracts towards the center of the triangle. The cost vertex also stays in place since the system was released as promised. Overall, the total impact on system value is less, as shown by the larger area in the triangle in Fig. 3.6c, but we have incurred technical debt.

Assuming we made the choice to incur the technical debt, what happens? The released version is not fully tested. Therefore, the system users may experience higher levels of system defects, affecting the usability of the system, and potentially their desire to keep using the system. If the tests are not executed, then additional problems

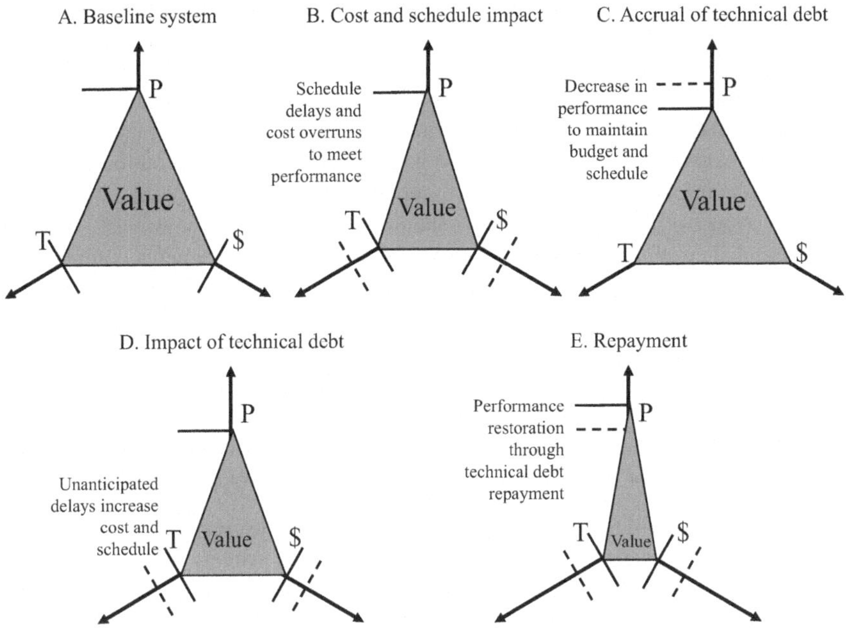

Fig. 3.6 Effect of technical debt on the system dimensions

with the design may not be revealed. It may take additional time to connect components that interface with the untested portions of the system, increasing the overall system development time. In the system dimension triangle, both the performance vertex and the schedule vertex contract towards the center, as shown in Fig. 3.6d, due to the presence of the technical debt.

The untested components of the system still need to be tested, and any deficiencies corrected to resolve the technical debt. This resolution takes additional time and schedule, potentially beyond the initial schedule and budget allocations, due to the interactions with additional components. This testing, and defect correction, will remove the technical debt from the system and restore the system performance, however, the overall value of the system is reduced, as shown in Fig. 3.6e.

The interconnectedness of the system dimensions shows the impacts of technical debt within the system. Technical compromises cannot be divorced from the associated schedule and budget impacts and short-term benefits to schedule or budget may carry long-term consequences in the performance dimension.

3.7 Technical Credit[2]

The discussion so far has largely focused on the negative impacts of decision making resulting in harm to the health of the system. However, the definition of technical debt states nothing about the type of impact. What happens if the impact on the system is positive? When focusing on the negative impacts, we have been talking about debts and fees, so a positive impact should be the opposite: a credit.

A technical credit is the inverse of technical debt and occurs when a compromise is made in any of the system dimensions to achieve a long-term benefit in the performance dimension. Examples include investing in automated tooling to improve the quality of manufactured parts and withholding a component from an iterative release to improve the delivered performance. Technical credits can be achieved when the system developer invests time or money to achieve an objective requirement instead of simply meeting the threshold requirement. Let's look at a more detailed example to see how a developer may choose to establish a technical credit.

Example The Hubble Space Telescope was designed to be serviced by astronauts while on-orbit. This capability was required to correct the primary mirror aberration, which was the result of technical debt (see Sect. 6.2). However, this capability also allowed for the accrual of several technical credits when additional repairs could be made on each servicing mission at low additional cost. A total of five servicing missions were performed on Hubble to both increase its capabilities and extend its lifetime [21]. Each of these missions had a primary objective, but the technical credits were realized when proactive upgrades were performed to increase the mission duration [22]. Each of these upgrades required additional investment by NASA in planning, training, and equipment and were not necessary to address the primary mission goals. However, these compromises in cost resulted in significantly increased performance.

This extra performance is the technical credit: by spending money, NASA increased the performance of the Hubble Space Telescope beyond the initial requirements. Technical credits can also apply to quality enhancements, process improvements, and other methods to decrease the potential for technical debt.

While the focus of this book is on technical debt in systems engineering, it is important to realize that possibilities for technical credit exist as well. A robust systems engineering management process (see Part III) will track both technical debt and technical credit, just like it tracks both risks and opportunities.

---

[2] The 'technical credit' was introduced by one of my colleagues, Nate Fikru. At an engineering review board, we were discussing a proposed system change that would improve the long-term performance of the system for a short-term cost. Since this change was the opposite of technical debt, he suggested that it we call it 'technical credit'.

3.8 Summary

This chapter provided an overview of technical debt, including its origins in software engineering and its transition to other fields, including systems engineering. The rise of Agile development methods across both software and systems engineering has opened projects up to increased levels of technical debt, as the focus of project delivery has changed from a single fully-functional product to rapid delivery of incremental improvements. These efforts produce different pressures on systems developers to make technical compromises to hit release dates, even if they may be able to implement a better solution.

This chapter concludes our introduction to technical debt and Part I of this book. In the next part, we define the causes and types of technical debt and discuss how they appear in the systems engineering lifecycle.

References

1. Uschold, M., & Jasper, R. (1999). A framework for understanding and classifying ontology applications. *Proceedings of the IJCAI-99 Workshop on Ontologies and Problem-Solving Methods (KRR5), 2.*
2. Kleinwaks, H., Batchelor, A., & Bradley, T. H. (2023, September). An Ontology for Technical Debt in Systems Engineering. *IEEE Open Journal of Systems Engineering, 1*, 111–122. https://doi.org/10.1109/OJSE.2023.3316395
3. Noguchi, R. (2024). The Nature of Technical Debt in the Development of Descriptive Models for MBSE. *INCOSE International Symposium, 34*(1), 516–526.
4. Li, Z., Avgeriou, P., & Liang, P. (2015). A systematic mapping study on technical debt and its management. *The Journal of Systems and Software, 101*, 193–220. https://doi.org/10.1016/j.jss.2014.12.0-27
5. Besker, T., Martini, A., & Bosch, J. (2017). Managing architectural technical debt: A unified model and systematic literature review. *The Journal of Systems and Software, 135*, 1–16. https://doi.org/10.1016/j.jss.2017.09.025
6. Lenarduzzi, V., & Fucci, D. (2019). Towards a Holistic Definition of Requirements Debt. *2019 ACM/IEEE International Symposium on Empirical Software Engineering and Measurement (ESEM)*, 1–5.
7. Ampatzoglou, A., Ampatzoglou, A., Chatzigeorgiou, A., & Avgeriou, P. (2015). The financial aspect of managing technical debt: A systematic literature review. *Information and Software Technology, 64*, 52–73. https://doi.org/10.1016/j.infsof.2015.04.001
8. Ampatzoglou, A., Mittas, N., Tsintzira, A.-A., Ampatzoglou, A., Arvanitou, E.-M., Chatzigeorgiou, A.,... & Angelis, L. (2020). Exploring the Relation between Technical Debt Principal and Interest: An Empirical Approach. *Information and Software Technology, 128*. https://doi.org/10.1016/j.infsof.2020.106391
9. Ciolkowski, M., Lenarduzzi, V., & Martini, A. (2021). 10 Years of Technical Debt Research and Practice: Past, Present, and Future. *IEEE Software, 38*(6), 24–29. 10/1109/MS.2021.3105625
10. Digkas, G., Ampatzoglou, A., Chatzigeorgiou, A., Avgeriou, P., & Matel, O. (2020). The Risk of Generating Technical Debt Interest: A Case Study. *SN Computer Science, 2*(1), 1–12. 10.1007/s42979-020-00406-6
11. Kruchten, P., Nord, R. L., & Ozkaya, I. (2012). Technical Debt: From Metaphor to Theory and Practice. *IEEE Software, 29*(6), 18–21.

12. Carson, R. S., Frenz, P. J., & O'Donnell, E. (2015). *Project Manager's Guide to Systems Engineering Measurement for Project Success: A Basic Introduction to Systems Engineering Measures for Use by Project Managers (Version 1.0).* INCOSE.
13. Kleinwaks, H. (2023). *Avoiding Technical Bankruptcy in System Development: A Process to Reduce the Risk of Accumulating Technical Debt.* Doctoral Dissertation, Colorado State University, Department of Systems Engineering.
14. McConnell, S. (2008, June). *Managing Technical Debt.* Construx Software Builders. Retrieved January 27, 2022 from https://www.construx.com/uploadedfiles/resources/whitepapers/Managing%20Technical%20Debt.pdf
15. Curtis, B., Sappidi, J., & Szynkarski, A. (2012). Estimating the Size, Cost, and Types of Technical Debt. *2012 Third International Workshop on Managing Technical Debt (MTD),* 49–53.
16. Kruchten, P., Nord, R. L., Ozkaya, I., & Falessi, D. (2013). Technical debt: towards a crisper definition report on the 4th international workshop on managing technical debt. *ACM SIGSOFT Software Engineering Notes, 38*(5), 51–54. https://doi.org/10.1145/2507288.2507326
17. SEBok Authors. (2024, November 25). *Risk (glossary).* SEBoK. Retrieved April 18, 2025, from https://sebokwiki.org/w/index.php?title=Risk_(glossary)&oldid=73237
18. Boswell, J. W., Anbari, F. T., & Via III, J. W. (2017). Systems Engineering and Project Management: Points of Intersection, Overlaps, and Tensions. *2017 Portland International Conference on Management of Engineering and Technology*, 1–6.
19. Walden, D. D., Roedler, G. J., Forsberg, K. J., Hamelin, R. D., & Shortell, T. M. (Eds.). (2015). *Systems Engineering Handbook* (4th ed.). John Wiley and Sons.
20. SEBoK Authors. (2024, November 25). *Quality (glossary).* SEBoK. Retrieved May 19, 2025, from https://sebokwiki.org/w/index.php?title=Quality_(glossary)&oldid=72807
21. National Aeronautics and Space Administration. (2025). *Astronaut Missions to Hubble.* National Aeronautics and Space Administration. Retrieved April 7, 2026 from https://science.nasa.gov/mission/hubble/observatory/missions-to-hubble/
22. National Academies of Sciences, Engineering, and Medicine. (2005). *Assessment of Options for Extending the Life of the Hubble Space Telescope: Final Report.* The National Academies Press. https://doi.org/10.17226/10374

Part II
Causes and Types of Technical Debt in the Systems Engineering Lifecycle

Chapter 4
Causes of Technical Debt

Abstract Technical debt can be intentionally introduced into a system as a response by the system developer to pressure from the system stakeholders. Examples are provided of how stakeholder pressure in each of the three system dimensions (schedule, cost, and performance) results in technical debt in a system. Additional causes of technical debt, including political pressure and technical trades are discussed and corresponding examples provided. Technical debt can also unintentionally be introduced into the system by the developers. Causes of unintentional technical debt, which is not introduced into the system as a direct response to external pressures, are provided.

Keywords Technical debt causes · Schedule pressure · Cost pressure · Technical trades · Political pressure · Compounding technical debt

In the previous chapter, we reviewed the types of technical debt present in the systems engineering lifecycle. This chapter moves the focus from the 'what' to the 'why'—why would a system developer take on technical debt? What motivations are there to make technical compromises? A 2023 study[1] attempted to answer these questions and the results are shown in Fig. 4.1. According to the results, technical debt occurs due to both intentional and unintentional decisions. These decisions are the result of schedule pressure, cost pressure, and technical trades.

In this chapter, we will review each of these causes of technical debt. Additionally, we will discuss three other causes of technical debt: political pressure, compounding technical debt, and unintentional technical debt. However, we begin this chapter by discussing how stakeholders generate pressure on system developers which results in the introduction of technical debt.

H. Kleinwaks, *Managing Technical Debt in Systems Engineering*,
https://doi.org/10.1007/978-3-032-30360-8_4

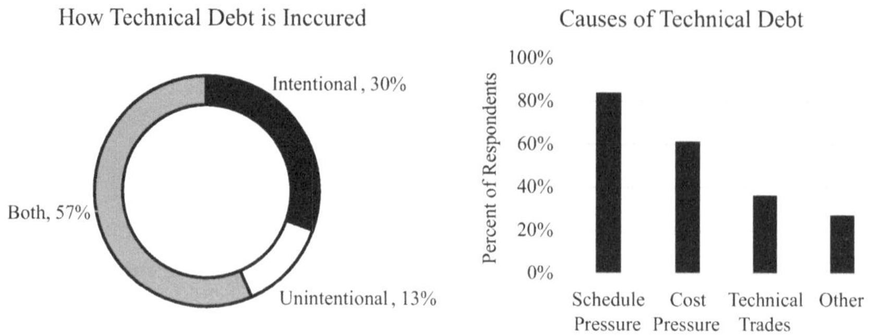

Fig. 4.1 Rationale for accruing technical debt [1]

4.1 Sources of Stakeholder Pressure

To understand how stakeholders create pressure on system developers, we need to return to the technical debt context map introduced in Chap. 3. The driving source is in the top left of the map (see Fig. 4.2): the stakeholders and their business goals. The system developer is beholden to system stakeholders, who often control the budget which enables the system to progress with its lifecycle. In order to satisfy these stakeholders, the developer may be enticed to take shortcuts in the system development process and to make technical compromises, even if these compromises are not the best choice for the long-term health of the system.

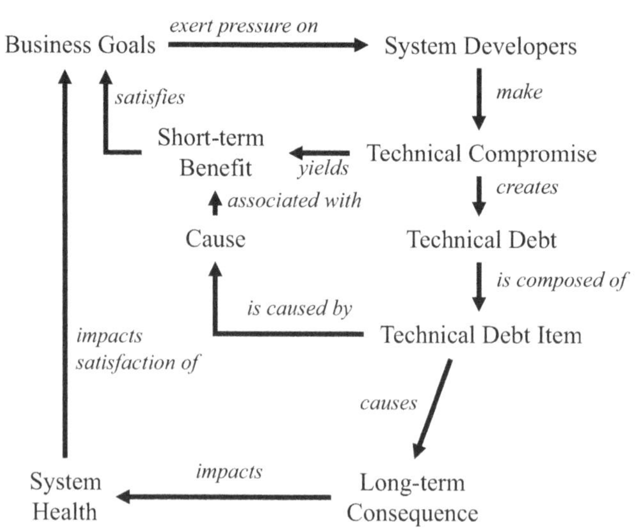

Fig. 4.2 Stakeholder pressure and business goals create technical debt

Why would the stakeholder put pressure on the system developer? It is helpful to think through the system dimensions and the value assessment from the stakeholder's perspective. While the system developer will feel the pressure to deliver value through the system performance, it is typically the stakeholder that will feel pressure to deliver on schedule and on budget, although they may convey this pressure to the system developer too! Therefore, their assessment of the relative value of each vertex of the triangle may be different from the system developer. The stakeholder may put extra value on delivering on time or on cost and less value on delivering the performance. External factors may be driving stakeholders, such as the need to publicize an on-time delivery and budget constraints. These factors produce outsized value for the stakeholder in the non-performance dimensions. This value assessment encourages the stakeholder to put pressure on the system developer, resulting in technical debt.

For example, a budget-conscious stakeholder may prefer the system shown in Fig. 4.3 B over that shown in Fig. 4.3 A. The profitability ($ vertex) is increased at the expense of performance (due to a higher selling point or cheaper development effort), which is more important to this stakeholder. Plotting the perception of what the stakeholder considers valuable on the system dimensions can yield insights into how they will pressure the system developer.

When dealing with stakeholders, it is important to consider their power over, and engagement with, the system. This relationship can be plotted on a matrix, as shown in Fig. 4.4 [2]. The matrix shows the power and engagement of the stakeholder in each quadrant along with the recommended stakeholder management strategy. Note that typical stakeholder matrices will use the term 'interest' instead of 'engagement.' However, since 'interest' is a key term in the technical debt ontology, we will use 'engagement' to indicate the stakeholder's involvement with the system to reduce confusion.

What does it mean for a stakeholder to have high power in the system? Power is a reflection of the level of control that the stakeholder has when it comes to system development. For example, on an internal research project, the Chief Financial Officer (CFO) may be the one who controls the research budget. Controlling the budget

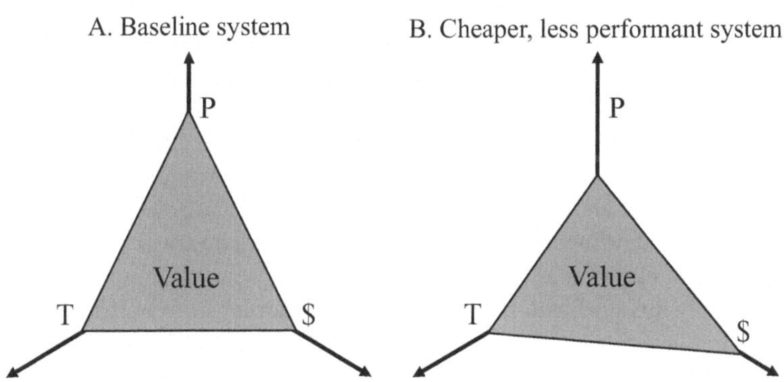

Fig. 4.3 Stakeholder view of system dimensions

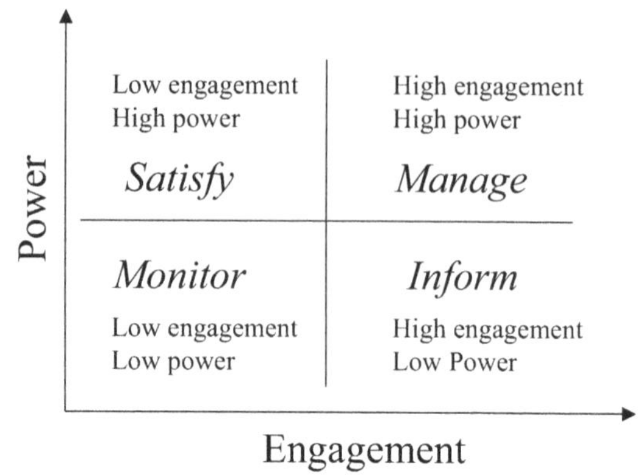

Fig. 4.4 Stakeholder power and engagement matrix

conveys power, as the CFO could put a stop to the project at any time if they deem the costs are no longer worth expending. Alternatively, the company Chief Technical Officer (CTO) may have high power when it comes to the technical performance of the system.

Engagement, on the other hand, reflects how much the stakeholder participates in the development of the system and cares about the performance outcomes of the system. High engagement indicates that the stakeholder is vested in the project outcome. The CTO may also have high engagement with the system and be monitoring the development closely, while the CFO may have low engagement and only be monitoring the budget through monthly reports. Users often fall into the high engagement, low power quadrant as they are very interested in how the system will perform and how it will be used, but have little say in the decisions that are being made during development.

The stakeholder power and engagement matrix includes recommended management strategies for each type of stakeholder. Those with high engagement and high power must be managed. To manage the stakeholder means to have frequent interactions to understand their inputs and to keep them informed of the progress of the system development, including letting them participate in review boards. Stakeholders with high power but low engagement must be satisfied. Satisfying the stakeholder means that they are kept aware of the figures of merit with which they are concerned (such as the budget or schedule) and that these figures of merit are within their expected ranges. Stakeholders with low power but high engagement should be kept informed—they are advised of progress and are offered opportunities to provide feedback. Finally, stakeholders with low engagement and low power should be monitored—the system developer needs to ensure that they are aware of progress and that their engagement or power is not likely to change.

If we overlay this grid with the technical debt quadrant from Fig. 2.2, we get the grid shown in Fig. 4.5, which maps stakeholders to the way in which technical debt is introduced into the system. The top right quadrant consists of the stakeholders who must be watched closely—those with high power and high engagement. Their high level of engagement indicates that they care about the success of the project and their

Fig. 4.5 Overlay of power and engagement grid and technical debt quadrant

High Power Low Engagement Negligent Debt	High Power High Engagement Strategic Debt
Low Power Low Engagement Unintentional Debt	Low Power High Engagement Tactical Debt

high level of power indicates that they have the ability to influence decisions. These stakeholders should be sources of strategic debt—technical debt taken on to advance project goals but managed to minimize the impacts on the long-term system health. In the bottom right quadrant, we have the high engagement, low power stakeholders. These stakeholders are sources of tactical debt—inadvertent but prudent debt. They come with suggestions for improvements and technical updates but are typically unaware of the associated budget and schedule implications. Responding to their requests can introduce additional scope, causing compromises to be made elsewhere in the system development. This introduction of debt is inadvertent since the stakeholder is unaware of the resulting trade-offs and prudent because the stakeholder's low level of power indicates that the changes can be addressed through a staged and managed process.

The stakeholders on the left side of the grid are those with low engagement in the system. On the bottom left are those with low power and low engagement; examples are vice presidents in different divisions that are not associated with an internal research project. These stakeholders would be the source of unintentional debt—the system developer is monitoring these stakeholders only and providing minimal responses to any suggestions. However, since these stakeholders are not highly engaged with the system, responses to their suggestions can result in unintentional debt—changes made to the system without considering the impact.

Finally, we have the top left quadrant, which represents the stakeholders with low engagement and high power and negligent debt—that which is introduced intentionally and recklessly. These stakeholders are not vested in the performance outcomes of the system development (low engagement), yet wield high power, such as a CFO with budget control. These are the stakeholders who are likely to demand that the product be released on time or that budget cuts be made, forcing the system developer to accept technical compromises that result in long-term impacts on the system.

Now that we understand the origins of stakeholder pressure, we can discuss the specific types of pressure that occur and the unique ways in which they create instances of technical debt. We start with the most common form: schedule pressure.

4.2 Schedule Pressure

Every project has deadlines, whether they be imposed by the stakeholders or self-imposed by the developer. If the project is behind schedule, then there is a choice at the deadline: delay the release of the system or release it as is, potentially with less performance than intended. Delays are often unpalatable to the stakeholder, and especially to high-power, low-engagement stakeholders whose primary concern may not be the system performance. Therefore, the stakeholder may pressure the developer to make technical compromises to ensure that the system releases on time, especially if the performance shortfalls can be hidden or delayed.

This pressure from stakeholders can occur in every release or delivery cycle and is exacerbated with frequent release cycles. If a Scrum development method is used, then the system developer is supposed to deliver a working version every sprint. Sprints represent fixed-time and fixed-price (assuming no workforce additions during the sprint) efforts and therefore the only variable is the delivered performance. The system developer has a choice if the development is behind: deliver fewer features (less scope) or make compromises (lower quality). Often, the stakeholder wants the full promised feature set, so the developer takes on technical debt by making technical compromises and delivering lower quality features.

These decisions are not unique to agile development methods—system developers face the similar decisions at gate reviews and milestones in every development method. However, they occur more frequently in agile methods, which increases the opportunities for technical debt insertion.

Schedule pressure can arise from multiple sources:

Demanding Stakeholder Similar to the example given above, demanding stakeholders insist on having system features delivered at a specific time. These demands could be based on the need to be first to market, to meet promised delivery dates, or to support other scheduled events. Demanding stakeholders are high in power on the power/interest grid and therefore need to be listened to, even if meeting their demands will limit system performance. Often, demanding stakeholders are more concerned about having a system delivered than the contents of the system. This type of schedule pressure can result in architecture debt, as the developer may take steps such as keeping prototypes or MVPs longer than desired.

Immovable Deadlines For some systems, there are truly immovable deadlines. For example, an interplanetary satellite has a defined launch time where the planetary alignment significantly shortens the journey. If this launch time is missed, then it may be several years before the alignment recurs, imposing an immovable deadline on the system. If the system is running behind, then technical compromises may need to be made to save schedule. These compromises often result in domain debt, where the delivered system does not function as intended in its domain, and test debt, where the system is not fully tested prior to release.

Insufficient Planning and False Promises Accurate schedules are difficult to create due to the unknowns that will occur during system development. However, a system developer needs to create a schedule that is feasible. Too often, system developers do not create a comprehensive plan. For example, a developer may not properly resource load their schedule and therefore not capture the effects of personnel availability on their ability to complete the project. Developers may also assume invalid learning curves during the creation of multiple components or incorrectly assume that they can parallelize tasks. Developers may not include enough margin within their schedule to account for the unknowns. This insufficient planning leads to false promises, especially in the proposal phase. The proposed schedule is often optimistic such that when the development starts, the schedule cannot be met. The stakeholder, however, was counting on the promised schedule and will attempt to hold the developer to it. These conditions result in the developer needing to make technical compromises to deliver a less performant system on schedule and are a form of requirements debt [3].

Supply Chain Delays A system typically is built from an interconnected set of components, often from many different companies. The set of parts, subsystems, and components is known as the supply chain. Delays in one part of the supply chain can cause delays in later parts of the supply chain, increasing the schedule pressure on the rest of the system development and introducing technical debt. Supply chain pressure can cause implementation debt, including substituting available parts for planned parts, and reducing quality controls such as performing less-thorough reviews of component test reports prior to integration into the system.

Cascading Delays in Late-Stage Processes Especially in sequential development methods, delays from earlier activities tend to compress later activities. Early activities are allowed to slip and extend since the deadlines are 'far away.' When late-stage system development activities occur, such as verification and validation, the time to delivery is short and there is increased pressure to finish quickly, resulting in technical compromises and test debt.

Example F-22 Rush to Production.

In 2001, 15 years after the program began, the F-22 aircraft program entered production. At this time, the program had only completed 26% of its drawings and only 44% of its manufacturing processes were under control. The development cost and schedule estimates exceeded the projections. Due to the immature state of the program, the production decision was forced to rely on more limited knowledge about the reliability of the aircraft [4].

This decision to enter production is an example of schedule pressure causing technical debt. In this case, limited reliability information (the technical compromise) was traded for starting production (the short-term benefit). The goal of the program, at this stage, was for the F-22 to demonstrate 2 hours of flying time between maintenance actions. However, the F-22 was only demonstrating an average of 0.44 flight hours between maintenance actions, a significant reduction in quality. In return for the benefit of starting production earlier, the system experienced the long-term consequence of significantly increased maintenance time and frequency and identified 260 different types of failures, all of which would need to be corrected, increasing the overall development time of the final system [4].

This example demonstrates how schedule pressure can produce test debt. Insufficient testing was conducted prior to making the production decision, resulting in increased failure rates and system maintenance. These consequences had the opposite effect to what was desired: instead of accelerating the delivery of functional aircraft, the increased failures increased program delays.

4.3 Cost Pressure

Budgetary constraints are a reality in system development. Even government projects, which are often maligned for excessive budgets, eventually face funding limitations or are subject to cancellation. For commercial systems, the amount of money spent on the system directly affects the profit (and return to shareholders) that can be derived from the system. Therefore, there is pressure on the system developer to stay within existing funding allocations and to limit the cost growth during development.

Cost pressure is not independent from schedule pressure, as schedule delays are often accompanied by cost delays. Schedule delays require more time from the people performing the work, which results in cost growth. Cost pressure forces the system developer to make a technical compromise to maintain schedule, thereby reducing costs. Similar to schedule pressure, cost pressure often comes from low engagement, high power stakeholders.

Cost growth is also triggered by changes to the system, especially those found late in the system development cycle [5]. Finding these changes late in testing can significantly impact both the cost and schedule of the system. These impacts are one of the reasons for the incorporation of agile methods and increased stakeholder involvement in systems engineering—frequent testing and feedback at lower levels decreases the cost and schedule impacts of any required changes.

Sources of cost pressure include:

Profit The profit is the amount of money that a company makes from selling its system. Without profit, the company will not be able to continue to grow. If the company is publicly traded, then profit is a key indicator of the company's success and impacts its stock price, which affects the shareholders. If a system is producing a low profit margin, then the company may insist on cost-cutting measures, which force the developer to make technical compromises, resulting in all forms of technical debt.

Labor Costs The labor costs of a system development refer to the salaries of the employees who perform the work. If there is pressure to limit the labor costs, then either fewer employees or less skilled employees will be performing the work. Therefore, less work will be accomplished. These impacts can manifest as documentation debt, design debt, modeling and simulation debt, and test debt.

Material Costs All systems require materials, tools, and supporting equipment during development. These support systems cost money, and cost pressure may result in using outdated equipment, inferior components, and insufficient tools. These compromises give rise to design debt, infrastructure debt, modeling and simulation debt, and test debt. As an example, spacecraft operate in an environment with increased radiation levels compared to Earth. The radiation levels affect computer systems, causing errors and resets. Computer parts that have been demonstrated to successfully operate in these environments are 'rad-hardened' and more expensive than non-rad-hardened counterparts. If faced with budget constraints, a spacecraft designer may select the cheaper, non-rad-hardened part, incurring design debt. This design debt is fee-bearing as the user of the spacecraft services will have to deal with increased outage rates due to radiation upsets.

Market Competition Systems that are sold to users face competition from similar systems built by other companies. The system stakeholders may insist on a sales price point to be competitive with the other companies, thus exerting cost pressure on the system developer to keep the budget to the point where the company makes a profit. This price point may not be based on the initial system budget, and therefore, the system developer may be forced into technical compromises to reduce the cost of developing the system.

Proposed Prices Similar to the sources of schedule pressure, a system developer may propose unrealistic budgets to secure work. Then, once the work has begun, they face challenges in delivering the proposed performance for the proposed price. If the customer is unwilling to increase the price, then the developer will need to make technical compromises to reduce the price [3].

Example Boeing 787 Dreamliner.

The Boeing 787 Dreamliner program provides an example of cost pressure leading to technical debt. During the development of the Dreamliner, Boeing implemented a risk sharing program with its suppliers where the suppliers were responsible for the design and fabrication of the parts and Boeing served as the integrator. In this arrangement, the suppliers bore the brunt of the upfront research and development costs and only saw a return on their investment once finished aircraft were sold. This risk-sharing arrangement substantially reduced Boeing's costs. Therefore, this arrangement can be seen as the result of cost-pressure on the Dreamliner development team [6].

Given that there was cost pressure on the development team, what technical compromise did they make? The risk-sharing approach was the technical compromise. Instead of using traditional and proven methods where Boeing would pay the non-recurring engineering costs to the subcontractors and would design the entire system, Boeing compromised the quality of the system by allowing the suppliers to create their own designs and retain the intellectual property [6]. Therefore, the suppliers were difficult to replace when they started exhibiting schedule delays. This is an example of design debt, specifically incompatible interfaces.

The design debt manifested with multiple consequences that negatively impacted the system health. Design issues included slower than expected integrated assembly at Boeing, structural flaws at the wing-body joint due to poor designs, and poor design communication from Boeing engineers to the suppliers. These issues alone caused over one year of delays to the first flight of the 787. Other technical debt, such as implementation debt, infrastructure debt, and documentation debt also contributed to the delays of the program [6].

It should be noted that the delays to the 787 were not exclusively due to the technical compromises. Some of the delays were due to Boeing programmatic decisions, such as delays in negotiating contracts with its suppliers. However, the cost-savings measures introduced a complex design structure that produced technical debt, which bore interest as evidenced by the increased complexity and delays accompanying the integration process.

4.4 Technical Trades

Modern systems are complex and often designed for multiple functions, and achieving optimal performance across all functions is not possible. Therefore, the system developer must execute trade-offs between design options, resulting in decreased performance across one or more functions and producing technical debt. In contrast to schedule and cost pressure, which are driven by stakeholders external to the system developer, technical trades often arise internal to the system developer as part of the design and implementation process. Technical trades are driven by high-engagement, high-power stakeholders.

Technical trades are also made to satisfy constraints on the system design. These constraints are often due to the environment in which the system operates. For example, the standard width of a shipping container is 8 ft [7], allowing it to fit on trucks, cargo ships, and trains. Therefore, any system component that is to be shipped by standard shipping container must fit within this constraint. If the developer is building a large structure, they need to make technical trades on the design to fit within this constraint. Options would include assembling structures on site after shipping, which could increase development time, reducing component size, which could reduce performance, or including folding mechanisms, which could reduce maintainability. These design decisions save the cost of customized, large-item shipping, but incur design debt.

Sources of technical trades include:

Environmental Constraints As discussed above, constraints of the system environment force the system developer to make technical trades. These trades result in design debt, when a less performant design must be selected in order to satisfy the constraint. Examples can include the corrosive effects of seawater, the radiation environment for spacecraft, or self-driving cars having to operate in adverse weather conditions.

Multiple Functions A system with multiple functions likely will not deliver optimal performance across all functions. Therefore, the developer must make trade-offs between the functions in the system design and implementation. Underperformance in a specific function, compared to user expectations, will result in technical fees.

Reuse of Proprietary/Existing Systems The results of technical trades are not always based on the best performing system. Other business considerations inform the trade analysis and can result in design debt and infrastructure debt through the selection of proprietary systems and interfaces, the use of COTS products, and the reuse of existing systems.

Components not Meeting Requirements A system is assembled from its components and if a component does not meet its requirements, then a technical trade may be required. The component can be replaced (causing schedule delays), the system design can be adjusted to accommodate the component, or the reduced performance can be accepted. These are examples of implementation debt.

Example Satellite development.

Technical trades often occur when there are hard constraints on the system. One example is the launching of satellites. Launching satellites is expensive and each additional kilogram (kg) of satellite mass increases launch costs. While recent commercial satellite launch companies have drastically reduced the cost/kg for launching spacecraft [8], each spacecraft must still meet a specified mass budget to achieve its desired orbit. This mass constraint forces satellite designers to make performance trades [9], such as:

- Limiting the size of the fuel tanks and therefore limiting satellite lifetime
- Limiting the size of the solar arrays and batteries and therefore limiting satellite payload performance
- Limiting communications system performance and therefore limiting the amount of data that can be collected and transmitted to users.

These trades represent technical debt: in order to meet the mass constraint, performance in other areas of the system is reduced. In these cases, the performance reduction will exhibit as technical fees with reduced user satisfaction.

4.5 Political Pressure

So far, we have discussed causes of technical debt that are driven by a stakeholder desire along one of the three system dimensions. However, pressure on a system developer can also be generated due to factors that do not align with a system dimension. We refer to this cause of technical debt as ‘political pressure’.

There are two different sources of political pressure— from stakeholders with high engagement and high power, and from stakeholders with high engagement but low power. High engagement and high-power stakeholders exert political pressure when they insist that their idea will produce a better solution, even if it results in a technical compromise elsewhere in the system. The stakeholder may believe that their solution will make the product more marketable, even if there is a reduction in performance. Because they have high power, the system developer will be forced to make the technical compromise.

Users are typically stakeholders with high engagement and low power. They are invested in the system development and care about the outcomes and will often provide feature requests and suggestions to the development team. These suggestions, when outside the system requirements, represent scope creep—the unauthorized addition of scope to the system. System developers may be pressured by the users to implement these out-of-scope features, especially systems with substantial user engagement during development. The pressure from the users is a form of political pressure, as they are not making the requests due to a constraint in one of the system dimensions. They are looking for a technical credit—a compromise in the cost or schedule dimension to increase the performance dimension. However, if not carefully controlled, diverting these resources may cause a compromise in another aspect of the performance dimension, creating technical debt instead. The prioritization of development tasks based on user-supplied priorities, which can disfavor infrastructure development, is a form of political pressure.

Sources of political pressure include:

Key Stakeholder Opinions Stakeholders with both power and interest will provide political pressure on the system developer to implement a specific design. The rationale for that design may not be related to schedule, budget, or technical trades and may instead be due to other factors. However, the development team may still need to make technical compromises to satisfy these influential stakeholders.

User interactions Users participating in the system development generally have a positive impact on the system development, ensuring that the system will meet their needs. However, users will request additional features that are beyond the requirements of the system. Because the developer is concerned about passing validation testing, they will feel pressure to implement these features, which may require technical compromises in other areas to balance the workload or deliver the requested feature on the new schedule.

Example Scottish Ferries Fiasco.

An example of political pressure causing technical debt is the Scottish Ferries Fiasco [10]. CalMac Ferris Ltd is a state-owned ferry service that provides essential transportation to 22 of Scotland's islands. In 2015, CalMac awarded a contract for £97 million to Ferguson Marine Engineering Ltd, a local shipbuilder, to build two new ferry ships. The first ship was supposed to be delivered in 2018, but completion dates for both ships delayed to 2024 and 2025 and the budget increased to over £360 million. Reporting by the BBC suggested that the selection of Ferguson Marine was based more on political rationale than on technical rationale [11].

Ferguson Marine submitted the most expensive bid, but was awarded the contract anyway. If the selection of an expensive, but qualified, contractor was the only issue, then this fiasco would not constitute technical debt. However, Ferguson Marine did not have experience building the types of boats required by CalMac [10]. The technical compromise made in this situation was compromising on the quality of the system by selecting an inexperienced contractor to award to a local company (the short-term, political, benefit). The long-term consequences include quality issues in the design, poor management, reduction in passenger capability, schedule delays, and budget overruns [10], eventually resulting in Ferguson Marine going bankrupt in 2019 and being nationalized by the Scottish Government [11].

The Scottish Ferries Fiasco shows how outside political considerations can push system developers into making decisions that incur significant technical debt. This pressure may be hard to avoid and may not be in accordance with proper practices but it still can result in significant technical compromises.

4.6 Compounding Technical Debt

The final cause of technical debt that we will examine is technical debt itself. Recall that the interest on technical debt consists of both the interest amount and the interest probability, where the probability is the likelihood that the developer will need to pay down the debt. Since this is a probabilistic quantity, that means that the technical debt may never need to be repaid! However, it also means that, if the debt does need to be repaid, the timing of when it will need to be repaid is uncertain.

Interest-bearing technical debt can hide in a system until a developer happens to attempt to modify or test that part of the system. Worse, during these times, other system components have interfaced with that debt-ridden part of the system. Now each of those components may need to be modified to fully remove the debt. This phenomenon is an example of compounding technical debt. Every part of the system that touches the debt-ridden part of the system increases the overall interest.

This compounding happens for two reasons. First, developers that interact with a component that contains technical debt may make design decisions that introduce additional technical debt. For example, if a system has documentation debt where the use of a component is poorly defined, the system testers may write test cases that do not reflect the proper system functions, introducing additional test debt to the system. If the testers were aware of the limitations of the documentation, they could write the correct test cases instead, containing the spread of technical debt. Second, correcting the original debt item requires a change to the associated artifact. All of the other artifacts connected to the original artifact would also need to be modified and/or regression tested. As more artifacts connect to the original artifact with workarounds, the taxes on the system increase since there are more artifacts that require corrections to the workarounds.

Sources of compounding technical debt include:

Lack of a Technical Debt Management Strategy Technical debt will be in many systems and is often unavoidable. That said, not all technical debt is bad, as long as it is properly managed. The lack of a technical debt management strategy means that the existing technical debt will remain hidden, resulting in compounding of the interest on that debt.

Loss of Institutional Knowledge Key personnel developing systems tend to know both how the system was designed and how it was actually implemented and where the implementation does not match the design. These personnel know where technical debt is in the system, even if it is not documented. However, if these people leave the program, they take this knowledge with them and any undocumented technical debt becomes hidden which leads to compounding interest.

Example Australian Stock Exchange Upgrade.

In 2016, ASX Ltd, the company which runs the Australian stock exchange, signed a contract with Digital Asset Holdings to update the platform used to make trades on the exchange to a blockchain based technology. Digital Asset was a startup which had demonstrated a test transaction to ASX. ASX bought a stake in Digital Asset and continued the project, despite concerns that the technology was not scalable and had not been tested [12]. In 2022, an Accenture report revealed that less than two thirds of software had been complete and that half of that software would need to be rewritten. The project was cancelled [13].

The failure of the blockchain upgrade can be seen as a result of compounding technical debt. The Accenture report [14] found the following issues in the program, among others:

- Insufficient tooling for test and release management (infrastructure debt)
- Misalignment between functional and non-functional requirements (requirements debt)
- Insufficient specification of scope (domain debt and requirements debt)
- Multiple technical issues that increase outages and functionality (implementation and design debt)
- Architectural problems that led to issues processing concurrent transactions (architecture debt)
- Significant gaps in the test management and quality management processes (test debt).

Each one of these issues was significant individually and played a part in the cancellation of the project. Compounding technical debt can be seen in two impact areas: the need to refactor code within the primary application and the effect on developers who connected their systems to the stock exchange. The Accenture report found that the planned refactoring work "would impact approximately 45% of the functional components that are already in testing. It would impose additional testing efforts for the changed components and additional testing phases (end-to-end, regression etc.)" [14]. By failing to address the technical debt early, Digital Asset Holdings was left with a major effort to fix the software.

The effects on the connected users, which manifest as fees, were also significant. ASX controlled the interactions with the Australian Stock Exchange. Therefore, any company that wanted to trade shares needed to interact with ASX's software. These companies were forced to deploy their own software engineers to keep up with the changes made by ASX, changes whose frequencies were influenced by the technical debt incurred in ASX's development process [12]. This impact on the downstream developers is an example of the technical debt compounding—a compromise related to insufficient testing in the core development product results in a release that is insufficiently tested. Developers adding new components adjust to that release, and then have to adjust again when the patch is released later, creating additional work throughout the interconnected system.

4.7 Unintentional Technical Debt

This chapter would be incomplete if we did not touch on unintentional technical debt. Technical debt may enter the system without a definitive cause—it cannot be traced to cost, schedule, or political pressure and it is not the result of a technical trade. This debt occurs because a system developer unintentionally makes a technical compromise, and they may not even be aware that they did so! For example, a systems engineer may write a requirement that is ambiguous. The engineer was not trying to write a poor requirement, but, nevertheless, they did so, and somehow the issues with the requirement were not caught through the peer review process. This unintentional action inserted requirements debt into the system.

Unintentional technical debt cannot be caught at the point of insertion, since there are no distinct events leading to a compromise that can be tracked. However, there are systems engineering practices that can reduce the likelihood of unintentional technical debt. Events like peer reviews and system design reviews can identify these issues and correct them, and are essential to minimizing the existence of unintentional technical debt. Techniques for identifying unintentional technical debt are explored in more detail in Chap. 8.

Sources of unintentional technical debt include:

Insufficiently Trained and Inexperienced Personnel Personnel working on the system without proper training or experience may introduce unintentional technical debt because they don't know any better. This type of technical debt is not introduced in response to outside pressures, but simply as a result of inexperience. For example, a full set of tests may not be performed on a component because the system developer did not know the tests should be performed.

Lack of a Technical Debt Management Strategy Unintentional technical debt occurs when the system developer is unaware of the long-term consequences of their actions. If a system does not have a technical debt management strategy in place, then there are no reinforcing mechanisms to develop the required awareness of technical debt to move its creation from unintentional to intentional. Part 3 of this book will provide the management strategy necessary to minimize occurrences of unintentional technical debt.

4.8 Summary

In this chapter, we examined the causes of technical debt. Stakeholders exert pressure on the system developer, which forces the developer to make technical compromises that result in technical debt. If we overlay the identified causes of technical debt, the technical debt quadrant, and the stakeholder power and engagement matrix, we get Fig. 4.6. Stakeholders with high power and low engagement will drive schedule and cost pressure, which can result in negligent debt. Stakeholders with high power and high engagement drive technical trades and political pressure, resulting in strategic debt. Stakeholders with low power and high engagement create political pressure and tactical debt, and stakeholders with low power and low engagement create unintentional debt. The goal of a technical debt management strategy is not to eliminate these sources of pressure on the system developer, but rather to enable the creation of strategic and tactical debt instead of unintentional and negligent debt. Therefore, extra attention must be paid to the high power, low engagement stakeholders to ensure they are not introducing negligent technical debt into the system.

In addition to the stakeholder pressure on the system developer, we also examined compounding technical debt and unintentional technical debt as causes of technical debt. Compounding debt occurs when technical debt in one area of the system causes technical debt in another area of the system. The tendency of technical debt to compound is what makes it especially dangerous in a system and reinforces the need to have a technical debt management process. Unintentional technical debt may be inserted into the system during nominal activities, and not necessarily in response to any direct stakeholder pressure. The system developer may not even know they are creating technical debt. Therefore, unintentional debt requires additional techniques to identify it before it creates problems (see Chap. 8).

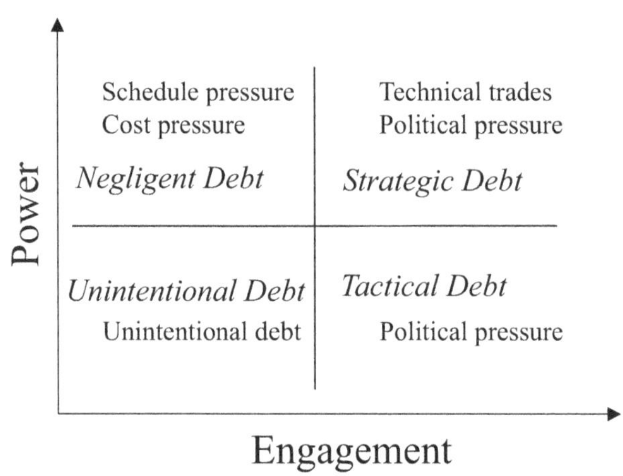

Fig. 4.6 Stakeholder power and engagement matrix with causes of technical debt and the technical debt quadrant

Now that we have defined the causes of technical debt, we need to look at similarities that will enable us to group together technical debt items to understand how technical debt appears within the system development lifecycle. In the next chapter, we discuss systems engineering-specific types of technical debt.

References

1. Kleinwaks, H., Batchelor, A., & Bradley, T. H. (2023). An Empirical Survey on the Prevalence of Technical Debt in Systems Engineering. *INCOSE International Symposium, 33*(1), 1640–1658. https://doi.org/10.1002/iis2.13104
2. Project Management Institute. (n.d.). *Stakeholder Analysis using the Power Interest Grid*. Project Management Institute. Retrieved 7 April, 2025, from https://www.projectmanagement.com/wikis/368897/stakeholder-analysis--using-the-power-interest-grid#_=_
3. Callister, P. S., & Andersson, J. (2016). Evaluation of System Integration and Qualification Strategies using the Technical Debt metaphor; a case study in Subsea System Development. *26th Annual INCOSE International Symposium, 26*, 1016–1028.
4. United States General Accounting Office. (2002). *Best Practices capturing Design and Manufacturing Knowledge Early Improves Acquisition Outcomes*. (GAO-02-701).
5. Walden, D. D., Roedler, G. J., Forsberg, K. J., Hamelin, R. D., & Shortell, T. M. (Eds.). (2015). *Systems Engineering Handbook* (4th ed.). John Wiley and Sons.
6. Zhao, Y. (2013). *Why 787 Slips Were Inevitable?* Rutgers, the State University of New Jersey. Retrieved 4 May, 2025, from http://zhao.rutgers.edu/787-paper-12-02-2013.pdf
7. Wikipedia contributors. (2025, April 8). *Intermodal container*. Wikipedia, The Free Encyclopedia. Retrieved April 8, 2025, from https://en.wikipedia.org/wiki/Intermodal_container
8. Jones, H. W. (2018). The Recent Large Reduction in Space Launch Cost. *48th International Conference on Environmental Systems*.
9. Oleson, G. (2014). *Relaxed Mass Limits: A New Option for Reducing Spacecraft Costs*. 30th Space Symposium. Retrieved from https://www.spacesymposium.org/wp-content/uploads/2017/10/GaryOleson_30th_Space_Symposium_Technical_Track.pdf
10. Calleam Consulting, Ltd. (n.d.). *Scottish Ferries Fiasco*. Calleam Consulting, Ltd. Retrieved April 8, 2025, from https://calleam.com/WTPF/?p=9404
11. Wikipedia contributors. (2025, April 3). *Ferry Fiasco (Scotland)*. Wikipedia, The Free Encyclopedia. Retrieved April 8, 2025, from https://en.wikipedia.org/wiki/Ferry_Fiasco_(Scotland)
12. Kaye, B. (2022, December 19). *Australian stock exchange's blockchain failure burns market trust*. Yahoo Finance. Retrieved May 17, 2025, from https://ca.finance.yahoo.com/news/australian-stock-exchanges-blockchain-failure-000538910.html?guccounter=1
13. Calleam Consulting, Ltd. (n.d.). *Australian Stock Exchange*. Calleam Consulting, Ltd. Retrieved May 17, 2025, from https://calleam.com/WTPF/?p=9512
14. Accenture. (2022). *ASX Chess Replacement Application Delivery Review*. Accenture. Retrieved 16 May, 2025, from https://www.euromoney.com/pdf/asx-chess-replacement-application-delivery-review-2022-pdf

Chapter 5
Types of Technical Debt

Abstract Many articles exist discussing technical debt, and different types of technical debt. However, the wide breadth of these discussions causes confusion—categorizations are too broad to be useful in developing strategies or too specific to be widely used. Therefore, a subset of technical debt types is required to ensure that management strategies can be developed for each one and properly applied. Within systems engineering, the types of technical debt are defined based on the system artifacts which that technical debt affects. This definition results in the following types of technical debt: architecture, design, documentation, domain, implementation, infrastructure, modeling and simulation, requirements, and test debt. Examples of each type of technical debt in real systems engineering projects are provided that identify the compromises made and artifacts affected.

Keywords Types of technical debt · Architecture debt · Design debt · Documentation debt · Domain debt · Implementation debt · Infrastructure debt · Modeling and simulation debt · Requirements debt · Test debt

By classifying technical debt into types, we can look for similarities between different technical debt items and derive common methodologies for identifying, managing, and mitigating the impacts of these items. When defining these types, it is necessary to strike a balance between overclassifying and under classifying. If we overclassify, then each instance could be a unique type of technical debt and developing common management strategies would be impossible. On the other hand, if we under classify, then we would try to apply common strategies to disparate types and the strategies would not succeed [1].

Therefore, this chapter defines a set of technical debt types that occur within systems engineering based on the system artifacts where the concession is made. This relationship is called out in the technical debt context map, as highlighted in Fig. 5.1.

Since technical debt requires a technical compromise, there must be a system artifact affected by the compromise [2]. By looking at the artifacts, common sources and compromises can be deduced leading to common management strategies. For

H. Kleinwaks, *Managing Technical Debt in Systems Engineering*,
https://doi.org/10.1007/978-3-032-30360-8_5

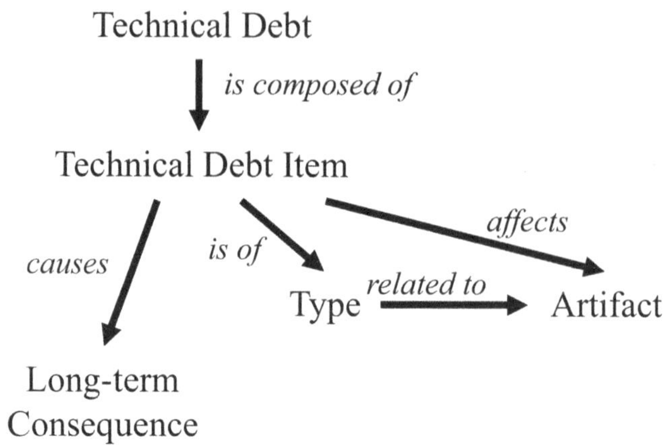

Fig. 5.1 Relationship of technical debt items, types, and artifacts

example, technical debt associated with the infrastructure required to support system development is grouped as infrastructure debt, regardless of the type of infrastructure. Similarly, all technical debt related to the ability of the system to be validated by its users is domain debt, since it affects the entire system. Table 5.1 provides the list of technical debt types in systems engineering that we will consider in this book. Each of these types, including the affected artifacts, will be defined in the following sections in this chapter.

In the following sections, we provide examples of each type of technical debt within actual systems development. Providing these examples is a challenge, since technical debt is not typically identified as a cause of systems issues. However, that does not mean that technical debt was not a problem. By reviewing reports from the United States Government Accountability Office (GAO), the "Catalogue of Catastrophe" from Calleam Consulting, Ltd. [4], and public news reports through the lens of technical debt, we can identify instances where technical debt was a likely culprit in the reported issues.

5.1 Architecture Debt

The system architecture defines the relationship between the system and its environment through the definition of the system operational concepts. The architecture is a high-level description focused on the structure of the system and the achievement of mission goals and is embodied in the system components, interfaces, and design. The system design establishes the detailed technical descriptions used to build the system that satisfies the architecture [5].

Technical compromises that impact the system architecture introduce architecture debt. Specific instances of architecture debt include incomplete interface control

Table 5.1 Technical debt types in systems engineering, adapted from [3]

Technical debt type	Example of artifacts where technical compromises are made	Interest or fee bearing
Architecture	ICDs and standards, concepts of operations	Interest and fee
Design	Design artifacts, including blueprints, system models, and algorithm descriptions	Interest and fee
Documentation	System documentation, including user manuals, training, and design description documents	Interest and fee
Domain	Statement of needs, system requirements, concept of operations	Fee
Implementation	Implemented and integrated system and its components	Interest and fee
Infrastructure	Support equipment: facilities, hardware and software tools and equipment, personnel	Interest
Modeling and simulation	Analytical and descriptive models, digital twins	Interest and fee
Requirements	Requirements specifications	Interest
Test	Test artifacts: plans, procedures, events	Interest and fee

documents (ICDs), relying on outdated standards and components, and failures to update the system operational concepts to reflect changes in the operational environment [3]. Research on architecture debt within software engineering produced the following five exemplars of architecture debt which are also applicable to systems engineering [6].

The Minimum Viable Product that Stuck In iterative development, the minimum viable product (MVP) represents the minimum set of functions in a system release to meet a given set of requirements or stakeholder needs. The MVP is the minimum work, not the complete set of work and may include items such as incompletely implemented ICDs and standards, incomplete testing, or missing documentation. However, the value produced by the MVP often creates excitement to move on to the next feature set, resulting in the acceptance of the MVP as the baseline architecture. Since the MVP never fully matured the system implementation, this acceptance can produce future complexities in system development when those missing implementations are required.

The Workaround that Stayed Often in system development, architectural workarounds are introduced, especially while waiting for components to reach maturity. Examples of such workarounds include temporary interfaces between components, use of non-standard interface options, and the use of non-standard components. Since these workarounds are functional and enable a system MVP, they end up becoming critical features of the system itself. Workarounds that were not part of the design often require the knowledge of the specific employees who implemented the workaround to make repairs or diagnose issues, exacerbating problems if those employees leave.

Re-inventing the Wheel Within systems development, the developer is often faced with a 'build vs buy' decision: do they create a new implementation of a component ('build') or do they reuse an existing component ('buy'). When the developer chooses to build their own version of the component, they are unable to leverage the existing testing and corrections that already exist in the component they could buy. While there may need to be changes unique to the components implementation within the system, the developer is also forced to re-certify compliance to all applicable standards and ICDs, which can increase development time.

Architectural Lock-in As a result of MVPs that stick and workarounds that stay, the system can develop architectural lock-in, where components become so deeply embedded in the system design that it is cost and time prohibitive to replace or upgrade them. The system architecture is dependent upon these out-of-date or incomplete components to execute, which can result in significant development costs or technical fees to the user in case of failure.

New Context, Old Architecture We like to design flexible systems that can operate in multiple different environments. However, that flexibility must be designed into the system from the beginning and old system architectures are not immediately adaptable to new system contexts. The failure to routinely update the system architecture, and any associated design elements, to keep pace with the evolving system context results in technical fees to the system user, as the system will not be as performant in the new context.

Architecture debt is interest-bearing. The reliance on outdated architecture components may increase the difficulty with which those systems are updated and with which new components are added to the system. Architecture issues have been shown to negatively impact the quality attributes of the system, including maintainability and evolvability [7]. Architecture debt is also fee-bearing. If the system architecture is brittle, then the operational system itself may be less reliable or less robust to failures. More frequent failures or failures that create longer outages will impact the users of the system.

Artifact(s):

- Operational concept documents
- System architecture description
- System interfaces, especially with external entities
- System context.

Compromises:

- Acceptance of MVP and workarounds
- Incomplete implementation of interface documents, especially with external entities

- Designing only for a single possible system context.

Impacts:

- Increased development to modernize out-of-date systems
- Increased risk of underperformance in different contexts or off-nominal situations

Example Southwest Airlines Scheduling Crisis 2022.

In the middle of the 2022 holiday season, Southwest Airlines cancelled more than 16,000 flights, costing the company over $800 million [8]. Severe winter weather disrupted travel from December 21 to December 25, which coincided with the busy holiday travel season. On the 25th and 26th of December, Southwest cancelled over 17 times more flights than Delta Air Lines, indicating that weather was not solely responsible for the Southwest issues. Why did Southwest have so many more problems than its competitors?

For years, Southwest had put off updates to its information technology (IT) infrastructure. Southwest's system was so bad that the pilot's union made upgrades to the IT system part of their bargaining demands and Southwest had to fall back on manual systems to schedule crew when outages occurred. Southwest's reliance on an outdated system prevented them from being able to promptly reassign crew to operate aircraft [9].

At issue in this example is the architecture of Southwest's system, which consists of the aircraft, crew, and computer systems. Southwest's system architecture consisted of a point-to-point route system, which prevented crew from being in a central location and prohibited easy rescheduling of cancelled flights [9]. This is an example of New Context, Old Architecture: the point-to-point architecture works within the nominal context. However, when the context changes to one of flight disruption, the old architecture failed. Their computer systems relied on an architecture of unreliable internal software systems and manual backups [9]. This architecture is an example of Architectural Lock-in: the old Southwest scheduling systems were locked into the foundational aspects of their overall IT enterprise and Southwest believed that it was not worth the cost to update.

These architectural aspects represent technical compromises: Southwest saved money by not upgrading their computer systems and flight structure at the expense of system performance. Therefore, this outage is an example of architecture debt. Every time that Southwest chose not to upgrade their systems and modernize their architecture, they achieved a cost savings benefit. However, each time they failed to modernize increased the likelihood of technical fees—in this case manifested as cancelled flights to the users. The 2022 holiday disruption was not the first time that Southwest had issues—their cancellation rate tripled between 2012 and 2022 and they cancelled more flights than their competitors [9]. Southwest's architecture was a major, and costly, source of technical debt.

Technical Compromise:

- Selecting to keep the old IT infrastructure in place.

Short-term Benefit:

- Cost savings related to not performing the update.

Long-term Consequences:

- Interest: system improvements limited by reliance on outdated architecture, new systems must work around inherent system limitations
- Taxes: multiple systems require upgrades at the same time
- Fees: extensive disruptions in case of system issues.

5.2 Design Debt

During design, the system developer translates the requirements into plans for building the system with sufficient detail that the system can be implemented [5]. The design of the system, often reviewed at both a preliminary design review and a critical design review, is developed and analyzed to ensure that the system will meet its requirements and operate in its intended environment. The design phase sets the baseline for the system that is then implemented, deployed, and operated.

During the design phase, the system developer must assess various options and execute technical trades. While a trade-off between design options does not represent technical debt in and of itself, if the results of the trade decision are made such that a lesser option with long-term impact to the health of the system is selected, then it could be technical debt. Throughout the design process, it is imperative to look for technical compromises due to insufficient analysis, misunderstanding of the environment, and a lack of focus on the 'ilities', such as maintainability and adaptability. Iterative design processes are particularly susceptible to design debt as initial designs may constrain the later design iterations [10].

Exemplars of design debt include:

Implementation Before Design Completion To get ahead in a project, or to make up lost time, steps that would normally be sequential are done in parallel. During the design phase, system developers will start implementing the system before the design is complete. This compromise, made to save schedule, carries the risk of future design changes affecting the already implemented part of the system, resulting in rework.

Misunderstanding of the System Context The system design must consider the requirements and the operating environment. Failure to consider both aspects can result in a system that fails validation. For example, a cell phone may be designed to function at room temperature. However, cell phones are rarely at room temperature, and if the phone shut down when exposed to the heat of sitting pool side in the summer then the user would experience a significant impact. Therefore, all the possible temperature conditions, and the full system context and environment, must be considered as part of the design.

Incompatible Interfaces The system developer must pay careful attention to the interfaces between components (both hardware and software) in the system. If the developer designs interfaces that are proprietary, exceptionally unique or complex, or generally incompatible, then the system will be more complicated to update and maintain in the future.

Unnecessary Complexity Designs for systems may often be more complex than required. Similar to the concept of reinventing the wheel in architecture debt, unnecessary complexity overcomplicates the design, making further system development more difficult. This complexity often arises from adherence to proprietary interfaces or prior work, instead of adapting simpler solutions from other sources.

Design debt is interest-bearing when the technical compromises impact the future development of the system. This occurs when designs make the system more complicated to implement or test. Design debt is fee-bearing when technical compromises affect the users, especially when the design does not support the entire operating environment.

Artifact(s):

- Design artifacts, such as computer drawings and blueprints.

Compromises:

- Use of proprietary interfaces instead of more common open-source interfaces
- Early start to implementation
- Selection of less-performant materials
- Designing without considering the 'ilities'.

Impacts:

- Longer integration times due to unique interfaces
- Reduced quality, maintainability, and availability
- Rework due to late design changes.

Example Apple's Resistance to USB-C Connections.

In 2022, the European Union passed a law declaring that all smartphones had to be equipped with USB-C chargers by the fall of 2024 [11], the culmination of a decade-long effort to enforce a common charging standard [12]. Although the European Union's efforts started in 2011, Apple introduced the proprietary Lightning port to its iPhone in 2012, setting up a challenge to the European Union's common charging standard goals. When the law passed in 2022, Apple was the only major phone maker not using the USB-C ports [11].

By using a proprietary port, Apple was able to control the entire ecosystem for its devices. Users had to buy Lightning adapters for equipment such as headphones and speakers [11] and were locked into compatible systems, whose makers licensed the technology from Apple [13]. This decision is an example of design debt—Apple decided to adopt a proprietary design instead of conforming to standards that might be codified in law. The short-term benefit was that they would boost revenues and sales. The potential long-term impact was the need to switch the design if the European Union law ever passed. Apple likely considered this technical debt as having a low interest probability at the time and that the short-term benefit of the additional revenue streams would be worth the potential consequences.

Certainly, the use of the Lightning port was a success for Apple. By some estimates, Apple will lose $5 billion per year or more through loss of revenues on Lightning cables and accessories [14]. Although Apple had to pay the interest on the technical debt in terms of a redesign of the iPhone to support the USB-C port, this example is a case where technical debt benefited the company. For ten years, they were able to make substantial profit off of a design decision. Although there are no explicit details on the cost to redesign the iPhone to include a USB-C port, Apple was already familiar with the technology through its use on other Apple products. Most of Apple's resistance to the change was based on claims of 'stifling innovation' and not on the technical ability to implement the change [11], and therefore we can conclude that this was not a significant technical hurdle for Apple to overcome. Additionally, Apple easily met the European Union's timelines, indicating that they may have been preparing for the outcome already. Apple eventually had to pay down the technical debt by adopting the common standard; however, the interest and principal were likely low enough to have made accepting the technical debt palatable. Through its use of USB-C on other products, Apple was managing this technical debt and mitigating its impact from a design perspective, leaving the only lasting impact in the lost revenue of Lightning accessory sales. Other systems that rely on proprietary interfaces may not be so fortunate.

Technical Compromise:

- Retaining Lightning port.

Short-term Benefit:

- Maintenance of revenue from Lightning-based accessories.

Long-term Consequences:

- Interest: continued development of Lightning-based accessories, all of which would eventually need to be redesigned
- Taxes: redesign of iPhones and accessories to support USB-C
- Fees: users had to continue to buy separate Apple and non-Apple charging cords and accessories.

5.3 Documentation Debt

System-level documentation is produced throughout system development. It is used to define the technical baseline, to define interfaces and user manuals, and to document design decisions. The migration to digital engineering processes has moved the documentation from traditional 'paper' based systems to integrated digital models that provide an authoritative source of truth. However, these models still represent the documentation of the system.

Critical to the successful management of any system documentation is both version control and configuration management. Version control refers to the management of changes throughout a specific document [15]. Version control is a component of configuration management. Configuration management is the process by which the configuration of the system baseline is controlled [5]. Configuration management is essential for ensuring that the system is always in a known state and that the baselined versions of components are compatible. The configuration must be documented and referenced to specific artifacts, such as particular versions of ICDs and standards. Any compromises made in the version or configuration control of the system, such as not using a version control system for critical documents, incur documentation debt.

Exemplars of documentation debt include [16]:

Insufficient or Incomplete Documentation If system documents have not been completed or do not provide a sufficient level of detail, then the system may incur documentation debt. Documents must be complete enough to enable their use. While knowing when to stop documenting may be more of an art than a science, it is better to have too much documentation than too little.

Outdated Documentation System documents must be kept up to date with changes to the system design, implementation, or the system context. Out-of-date documentation provides an illusion of sufficiency; however, interest on the debt occurs when the developer must first decipher whether or not the documentation is relevant to the task at hand. Outdated user manuals or maintenance instructions create fees for the users and may render the documents unusable.

Incorrect Documentation In addition to insufficient and outdated documentation, documentation debt occurs with documentation that is incorrect. Errors in the system documents result in incorrect system development activities that can be time-consuming to identify and fix. Erroneous specifications can result in a system that is built to its specification yet still does not meet its performance requirements.

Poor Configuration Management Processes Configuration management ensures that every system developer is using the correct version of the system documentation when performing their design. The lack of proper configuration management can result in developers designing to outdated interface specifications, requirements, or component designs. Using outdated documentation will produce incompatible products which will need to be corrected later in the development cycle when the components are integrated.

Documentation debt is interest-bearing when the documentation pertains to the development of the system. Insufficient design descriptions, code comments, and other information can lead to increased complexity of system updates and deficiency corrections. Documentation debt is fee-bearing when the documentation pertains to the interaction of the system and its users, such as lack of training and user manuals.

Artifact(s):

- All documents and models used to design and maintain the system and its components [17].

Compromises:

- Not updating documentation until the end of the program, leading to out-of-date specifications during system development
- Foregoing formal ICD updates when changes are necessary during integration
- Updating documentation without releasing it into the configuration management process results in different components using different versions.

Impacts:

- Increased difficulty updating system components due to lack of proper documentation
- Increased difficulty using the system due to insufficient training and user manual
- Incompatibility between components due to loss of version control and configuration management.

Example Airbus A380 Production Delays.

The Airbus A380 is a massive aircraft, designed to carry a maximum load of 850 passengers. The initial production of the aircraft experienced delays due to problems with the 530 km of wiring, concurrent design and production, and issues with change control and configuration management [18].

Airbus is a multinational corporation with divisions across different European countries. While all offices at Airbus used the CATIA CAD software, the German and Spanish designers used version 4 and the British and French designers used version 5. Critically, the German designers worked on the electrical system and the French designers developed the airframe. The designs were integrated into a single 3D Digital Mock-Up which revealed no problems. The problem existed in the upgrades between the software versions: CATIA version 5 had updated the bend radius calculations for wires, and when there are 530 km of wires, the differences add up [19]. Additional configuration management issues included the use of different conductors for the wire harnesses and the use of non-standard dimensions [18]. The end result was that the wires were too short and would not fit in the airframe. The different CAD programs created incompatible design documents resulting in erroneous specifications. While this issue could be considered infrastructure debt due to the use of COTS software, we prefer to characterize this issue as documentation debt due to the lack of proper configuration management. Because the configuration between the two different CAD programs was different, each team worked on different assumptions and designed to invalid interfaces.

Technical Compromise:

- Use of dissimilar design processes and tools without proper configuration management.

Short-term Benefit:

- Acceleration of design at each site due to familiarity with tools and processes.

Long-term Consequences:

- Interest: Creation of custom file conversion tools [18] and incorporation of incorrect specifications into drawings and designs
- Taxes: Significant extra work required to correct issues with wiring, resulting in billion-dollar delays [19].

5.4 Domain Debt

Often in systems engineering, we talk about the difference between verification and validation. Verification determines if we 'built the system right,' in accordance with the requirements, while validation determines if we 'built the right system,' as determined by the system users. Validation, by its nature, is more subjective than verification and deficiencies in validation may be harder to fix. Iterative methods seek to minimize the risk of failing validation by encouraging early interactions with users during the development phases.

Domain debt is the "misrepresentation of the application domain by an actual system" [20] and relates to failed validation. Technical concessions during the development process lead to long-term impacts on the usability of the system by its users. The 'misrepresentation of the application domain' refers to conditions where the system was not 'built right'—it does not properly interact with its intended environment or does not provide the feature set needed by its users. Systems which experience domain debt may not be deemed suitable for their intended use.

Exemplars of domain debt include:

Misunderstanding of Requirements A system developer may have a different interpretation of the system requirements from the system stakeholders. The developer will proceed to develop a system that meets their interpretation of the requirements and which may be able to pass verification. However, since the stakeholder's interpretation of the requirement is different, the system will fail validation.

Inflexibility to Change Unfortunately, systems are not developed in vacuums and the system context and use cases may change during system development. If the system developers choose not to respond to these changes, then they may incur domain debt as the user expects the system to have the same performance within the changed system context.

Domain debt is usually fee-bearing. The misunderstanding of the system context results in issues with system performance and validation. While rework of the system will be required to remove the domain debt, the long-term consequences of domain debt are reductions in system utility and performance for the user.

Artifact(s):

- The system itself as expressed through key system documents such as statements of needs, requirements, and concepts of operations.

Compromises:

- Limited user engagement
- Failures to update system components and interfaces
- Reliance on old technologies and systems.

Impacts:

- Failure of validation due to system not meeting user needs
- Insufficient system performance, especially in off-nominal cases.

Example Avon Service Model Transformation.

Avon is a well-known door-to-door seller of makeup products whose business model relies heavily on direct sales from their representatives, who often sell the makeup products to friends and family. Without these sales representatives, Avon would lose its primary distribution channel. Therefore, it would make sense for Avon to develop processes and systems to enable their sales representatives to be more efficient and to sell more products [21].

In 2013, Avon introduced the "Service Model Transformation" project, which included a tablet-enabled ordering system allowing sales representatives to display and order products for their customers directly from the tablet. Unfortunately, the interface developed on the tablets was difficult for the sales representatives to use. Design issues significantly degraded the user experience. The difficulties were substantial enough that sales representatives stopped working for Avon, with one senior executive reporting a loss of 1/3 of her sales team. The project was supposed to save Avon $40 million per year but ended up being canceled with a write-off of over $100 million [21].

This project is an example of domain debt. The system was supposed to increase the efficiency of sales representatives and enable them to sell more products. Even though Avon limited the roll-out of the system to Canada, the developers did not interact with users (sales representatives) enough to understand how the system needed to function in its intended environment. Representatives had difficulties with even the basic functions, such as logging in and accepting orders. Avon claimed that the technology worked well but sales representatives disagreed—a clear indication of the misalignment between the system developer and the application domain [22].

In this case, the technical compromise is hard to ascertain, but it is clear that user feedback was not well integrated into the system. The long-term impacts on the system health included the difficulty in processing orders, dropping orders, and the eventual loss of sales representatives.

Technical Compromise:

- Limited testing and trials with sales representatives.

Short-term Benefit:

- Delivery of the "Service Model Transformation" project.

Long-term Consequences:

- Fees: system so difficult to use that some sales representatives quit instead of continuing to use the system.

5.5 Implementation Debt

The implementation phase is when the system moves from a design on paper to an actual physical system. In sequential development lifecycles this phase starts after the critical design review while in adaptive lifecycles this phase occurs in each iteration after the design for that iteration is complete. During this phase, the system developer has to interpret the design to turn it into reality, and these interpretations require the developer to make choices on the implementation. These choices do not affect the system architecture or the design, but different selections can impact the resulting system performance. Implementation debt is technical debt that is accrued due to these choices.

Exemplars of implementation debt include:

Inefficient Implementation There may be various ways to implement a system that is compliant with both the system design and the system requirements. In software systems, inefficient implementation of the designed algorithms can cause performance issues due to memory problems or increased processing times. This issue is an implementation issue and not a design issue, since it relates to how the design is translated into the actual system.

Lack of Quality Controls and Processes Louis Gertsner, former CEO of IBM, once said "People don't do what you expect but what you inspect" [23]. Within systems engineering projects, the quality assurance group is responsible for ensuring that the project adheres to the design processes while quality control ensures that the produced products meet the required standards [5]. However, quality assurance and quality control can be expensive to implement. Concessions in either group may save money but will also reduce the quality of the finished system product, creating technical debt.

Implementation debt is interest-bearing when the concessions made during system implementation make future system development more challenging. For example, if a lack of quality controls results in components that are not regularly within tolerances, then integration of these components will take longer than expected. Implementation debt can also be fee-bearing when the concessions made during implementation produce a less-performant system, such as one that requires additional unplanned maintenance.

Artifact(s):

- System components and integrated system.

Compromises:

- Reduced quality control programs and testing / inspections
- Reduced peer reviews
- Reduced training of system developers.

Impacts:

- More frequent rework requirements
- More time-consuming integration across components
- Reduced performance due to system quality issues.

Example Samsung Galaxy Note 7 Battery Explosions.

Samsung released the Galaxy Note 7 mobile phone on August 19, 2016 to high demand, with preorders breaking records. On September 15, 2016, Samsung announced a formal recall of the phones due to a manufacturing defect that caused the battery to explode. New phones were created with a different battery, from a different supplier, however the new phones were subject to similar explosive incidents, leading Samsung to cancel production of the phone in October 2016, less than two months after the official release [24].

The first set of phones had a design flaw in the battery that resulted in the combustion of the phone. The second set of batteries exhibited explosive tendencies due to a manufacturing defect [25], and it is this second set that was subject to technical debt. In this case, Samsung was under schedule pressure to correct the battery problem on the newly-released phone to maintain sales and deliver working products to their customers. Believing that the battery issues were a design flaw, Samsung accelerated the production of the second battery, from a different company. This rush to production impacted the quality of the battery production and resulted in "unknown manufacturing issues." In the time since, Samsung has instituted additional quality control measures [26].

The battery issue from the second manufacturer is a clear case of technical debt, albeit one with a consequence that was rapidly observed. In this instance, the battery manufacturer did not implement sufficient quality controls to meet the high schedule demand, resulting in implementation debt. The long-term consequence must be viewed from the perspective of Samsung as a whole and not just the specific product. Samsung suffered damage to its reputation, lawsuits, and financial losses as a result of the decisions to rush the secondary battery into production. Additionally, the users of the Galaxy Note 7 experienced significant fees—not being able to bring their phones on airplanes, having to return their phones, and the ever-present risk of an explosion [24].

Technical Compromise:

- Rushing secondary battery through production without proper quality controls.

Short-term Benefit:

- Delivery of battery replacements for troubled Galaxy Note 7 phones.

Long-term Consequences:

- Taxes: implementation of increased quality controls in future productions, reputation damage
- Fees: users return phones, users unable to travel with phones, phones explode.

5.6 Infrastructure Debt

During system development, we have a tendency to focus on the system that is being developed. Underlying that system is the infrastructure: the set of tools, facilities, and people that support its development, deployment, use, and retirement. The system infrastructure consists of various items such as the computer systems and software used to develop software code, specialized and general test equipment, and the factories used to manufacture parts.

Technical compromises made in the configuration of the system infrastructure result in infrastructure debt.

Exemplars of infrastructure debt include:

Commercial-Off-The-Shelf Products Using Commercial-Off-The-Shelf (COTS) products can provide quick value to a system developer. Instead of building the same capability (and potentially being subject to Architecture Debt), the developer acquires the capability through purchasing existing products. However, the system developer must be careful in these situations as the COTS product may not be easily modified to fit the needs of the system. COTS products may only have some of the required capability, forcing the system developer to create custom interfaces and functions to deliver complete system functionality. If the actual capability set of COTS products is not well understood or the COTS product is expected to be more performant than it is, then COTS products can result in lower overall performance and can complicate system integration through the need to develop customized interfaces [27].

Automated Build Pipelines In software systems and subsystems, automated build technologies enable rapid deployment and upgrade of software versions in automated fashion. The lack of these pipelines can cause infrastructure debt as the resulting software deployment efforts will take longer and be more error prone. Additional attention must be paid to ensure that the correct tools are in the build pipeline: if the proper cybersecurity scanning tools are not included, then the resulting system might not be cyber-secure.

Depreciation of Infrastructure Infrastructure components, especially facilities and test equipment, that are allowed to depreciate can negatively impact the development of the system. Older equipment is often less performant, increasing the time it takes to execute the tasks. This equipment is also more prone to failure, increasing facility downtime and extending system development time.

Insufficient Staffing System infrastructure includes the people who do the work of developing the system. System development projects that are understaffed or utilize poorly trained personnel are subject to infrastructure debt—the development will take longer due to the lack of workers.

Infrastructure debt is interest-bearing as infrastructure issues tend to affect the system development timelines. The supporting elements used to create the system are not used in support of the deployed system. While the final system may be less performant due to infrastructure choices, the primary impact is on the system development.

Artifact(s):

- Equipment, tools, and personnel supporting the system development.

Compromises:

- Using COTS products that are not a good-enough fit
- Insufficient supporting equipment, either in amount or performance
- Letting supporting equipment age past its useful lifetime
- Understaffing or undertraining personnel.

Impacts:

- Increased downtime due to infrastructure maintenance
- Increased development time to develop custom interfaces for COTS products
- Increased schedule due to lack of personnel.

Example Panorama.

The Panorama project was an IT system developed in Canada intended to help manage pandemics. It was initiated in response to the 2003 SARS (Severe Acute Respiratory Syndrome) outbreak and was awarded to IBM in 2006 [28]. Panorama was designed to create a system to allow coordination between public health officials in the case of a larger disease outbreak. Unfortunately, the system was plagued by problems, as identified in a 2015 audit report. During the COVID-19 pandemic in 2020 only some parts of the Panorama system were usable [29].

The Panorama system was accepted in 2010, with 1200 defects. Between the time the system was deployed and the time of the audit, the defect number increased to over 11,000! The system suffered from usability issues, slow performance, and frequent outages, meaning that the Panorama system decreased the efficiency of the

Canadian health system. These impacts are an example of technical fees: longer wait times, cost increases, and reduced capacity affect the users (the patients and the medical professionals) every time the system is used. By 2015 the project was 420% over budget [28].

One of the reasons cited in the audit report for why the system failed was the plan to combine multiple pieces of COTS products into the final solution. The project requirements specified that COTS products needed to be integrated with new software as previous experience had shown this approach to be cost-effective. However, in this area of public health, there were no existing COTS products that met all of the requirements. Therefore, the system developer, IBM, proposed to combine multiple pieces of software together to create a single, country-wide system. IBM underestimated the complexity of this process due to the different needs of the separate Canadian jurisdictions, each of which has different processes, languages, and population sizes. In 2007, the system development passed the planned completion date of the national system without a product and IBM replanned the development, shifting to a custom solution instead of the COTS solution, receiving ~ $10 million in additional funding and pushing the scheduled delivery date of the national system to 2010 [28].

The Panorama project is an example of infrastructure debt through the selection of COTS software. The system stakeholders desired to save schedule and budget and therefore compromised the performance of the system by relying too heavily on COTS software that did not meet the needs of the project. Proper diligence on the part of the system developer and the stakeholder would have revealed the complexities of integrating these various software components and either properly budgeted for them or selected a different approach. Instead, they selected existing software which was insufficient to meet the national needs of the Panorama system. The result delayed the deployment of the system. The principal and taxes were eventually repaid, costing an additional $10 million and three years of schedule.

Technical Compromise:

- Using COTS software without fully vetting capabilities.

Short-term Benefit:

- Accelerated schedule based on assumed COTS capabilities.

Long-term Consequences:

- Interest: Increased cost and rework of the system to work around the failure of COTS modules
- Taxes: Eventual replacement of the COTS modules with a custom solution.

5.7 Modeling and Simulation Debt

In modern systems development, computer-based models and simulations are frequently used to evaluate system designs and expected performance. These digital models and simulations take the place of expensive prototyping and testing and enable the rapid assessment of multiple design candidates in various operating environments. Models and simulations are used early in the system lifecycle, especially in the concept development and early design phases, to evaluate different system architectures and potential designs.

While modeling and simulation is part of the design of the system, the use of these tools to evaluate designs puts them into their own category of technical debt. The models and simulations must be developed to the appropriate levels of fidelity to assess system and component design and to predict performance. If there are shortcomings in the modeling and simulation tools used to create requirements, then even if the system is designed and implemented properly, it still will not perform as expected when deployed. For example, the orbit of a satellite is affected by many different forces, such as the gravity of the Earth, sun, and moon, atmospheric drag, and solar wind. If an Earth-imaging satellite is being designed and the orbital analysis does not account for the atmospheric drag, then the satellite will experience a shorter lifetime than desired and predicted by the model. The lack of modeling fidelity is an example of modeling and simulation debt.

Exemplars of modeling and simulation debt include:

Insufficient Fidelity The fidelity of a model and simulation refers to the accuracy with which the system and its operational environment are modeled. Typically, increased fidelity in the models and simulations increases both the cost of the work and the runtime of the system. Therefore, system developers try to limit the fidelity to the appropriate levels for the system they are assessing. However, insufficient fidelity can result in overly optimistic performance predictions and invalid system designs. These errors may not be found until the system is under test or in operations, when the cost to fix the errors rises significantly.

Invalid Assumptions No model can capture the entirety of the system behavior and modelers will make assumptions about the system, its environments, or both. Often these assumptions are based on outputs from other models and used to decrease the runtime of the current simulation. Sometimes, however, these assumptions are due to a misunderstanding of the system requirements or its operational context. If the design of the system hinges on critical assumptions that prove to be invalid, then redesign and retesting may be required or performance degradation may occur.

Invalid Figures of Merit The system is evaluated against figures of merit—specific analytical assessments that highlight parameters critical to the success of a system. For an electric car, figures of merit may include items such as the charging time and the range on a full charge. Invalid figures of merit are those that do not properly capture the needs of the stakeholders. For example, if the stakeholders of the electric car are most interested in the charging time and the figure of merit is battery capacity, then the system would be designed to maximize the battery capacity and result in a longer charging time. Because the right figures of merit were not modeled, the performance of the final product would be reduced.

Modeling and simulation debt is interest-bearing when the deficiencies in the models are found during system implementation and testing. These deficiencies will cause rework and redesign, increasing overall system development time. Modeling and simulation debt is fee-bearing when it incorrectly predicts performance, reducing the value that the user receives from the system.

Artifact(s):

- Models and simulations used to predict and evaluate system performance.

Compromises:

- Reduced fidelity in the models and simulations
- Unverified assumptions and figures of merit.

Impacts:

- Reduced performance in actual operating environments
- Failed testing requiring rework.

Example Mars Climate Observer.

The Mars Climate Observer is infamous for crashing into Mars due to a unit conversion error. While the developers of a software routine to calculate small forces on the spacecraft did provide the results in English units instead of in metric units, the use of inconsistent units was not the only problem with the spacecraft modeling and simulation. James Oberg provides a great overview in his IEEE Spectrum article "Why the Mars Probe Went Off Course" [30]. In this article he identifies multiple instances of technical debt, including modeling and simulation debt.

The Mars Climate Observer was asymmetric with a single solar array. This single array caused an uneven distribution of mass and produced an uneven disturbance torque on the spacecraft, requiring careful management of spacecraft momentum to prevent it from tumbling out of control. To counteract the unbalanced torque, the NASA designers intended to spin the spacecraft along the axis perpendicular to the sun. However, this spinning motion would reduce the sunlight on the solar panel, which raised concerns about the available power. Therefore, the design team changed the orientation such that the solar array would always face the sun. Critically, there were no navigation experts on the design team, so, presumably, accurate modeling and simulation of the effects of the orientation change on the ability of the spacecraft to navigate were not well understood, thus incurring modeling and simulation debt.

The positioning of the solar array resulted in an increased use of the Mars Climate Observer's thrusters to dump angular momentum from the reaction wheels. Each time that the thrusters were fired, the units error introduced incorrect forces into the navigation software. Eventually, the difference between the modeled and actual trajectory was so great that the spacecraft ended up crashing into the Mars atmosphere.

Had NASA included navigators earlier in the development process, the models and simulations used to develop and validate the software, the attitude design changes, and the in-flight models themselves may have come under better scrutiny. Instead, the Mars Climate Observer team took on the technical debt from not performing sufficient modeling and verification. Unfortunately, the interest on this debt was realized in an extreme way.

Technical Compromise:

- Change in the solar array orientation without consulting navigation experts creating a deficiency in the spacecraft models and simulations.

Short-term Benefit:

- Increased power profile for the satellite.

Long-term Consequences:

- Interest: Increased use of thrusters to work around asymmetric torque, which magnified errors in the navigation software
- Fees: Loss of spacecraft.

5.8 Requirements Debt

The system requirements are the specifications that define how the system is to be built. They represent the recipe by which the stakeholders' needs are turned into the system by the developers. Getting the requirements correct is such a critical aspect of systems engineering that entire tools (such as DOORS[1]), products (requirements verification traceability matrices), gate reviews (system requirements review (SRR)), and test efforts (verification ensures that the system is built to the requirements) are centered on the system requirements.

There are differences in opinion about whether requirements debt should be considered a unique form of technical debt [32]. After all, concessions made in verifying the requirements would be test debt, compromises made in the development of the requirements that led to a misunderstanding of the user needs would be domain debt, and poorly formatted requirements documents could be considered documentation debt. However, given the importance of well-formed requirements to the successful execution of a system, requirements debt must be considered a separate type of technical debt, with the focus on compromises made in the development of the requirements themselves.

System requirements must follow the quality attributes defined in Table 5.2. Requirements debt occurs when the requirements writer does not take the time to follow these requirements guidelines, resulting in missing, conflicting, ambiguous,

[1] Dynamic Object-Oriented Requirements System (DOORS) is an engineering requirements management tool provided by IBM. DOORS provides the ability to link requirements to each other, to design items, to test cases, and to test plans [31].

unnecessary or untestable requirements. These requirements can have significant impacts on the system development and implementation, causing unnecessary swirl during the design and test phases.

Exemplars of requirements debt include:

Missing Requirements Requirements writers may underspecify a system, leaving the system developer to guess at the desired implementation. If the developer guesses wrong, then they will be subject to rework, which may not be caught until late in the test phase. Missing requirements are a source of validation failures and expensive changes to the design. Missing requirements may occur due to the failure to include all relevant stakeholders in the requirements generation process.

Ambiguous Requirements Similar to missing requirements, ambiguous requirements can result in an undesired system as the developer must guess at the proper interpretation. Ambiguous requirements occur when proper care is not taken in the writing and review of the requirements.

Inadequate Requirements Reviews Most system lifecycle models contain requirements review gates. At these reviews, it is critical that the system developer and stakeholders review and agree upon all requirements and the methods by which those requirements will be verified. This agreement sets in motion the rest of the system development, as the requirements will inform the design, which informs the performance. If all stakeholders, including operators, are not represented at the requirements reviews then the risk of missing or ambiguous requirements increases.

Table 5.2 Requirement quality attributes [33]

Quality attribute	Description
Complete	The requirement contains all relevant information, such as version number, aspect of the system affected, and due date
Traceable	The requirement contains traceability information to its source and previous versions to enable the system developer to understand its origins
Correct	The requirement contains the proper specifications such that, when satisfied, the system realizes the expected performance, without requiring additional capability
Unambiguous	The requirement is written such that there is only one possible interpretation
Comprehensible	The requirement is easy to understand
Consistent	The requirement does not contain statements which contradict themselves or other requirements of the system
Verifiable	The requirement is able to be verified through an acceptable verification method
Atomic	The requirement defines a single requirement on the system and not a compound set of requirements
Up to date	The requirement represents the current specification of the system

Requirements Changes and Scope Creep Changes to the requirements, either modifications to existing requirements or additions to the system scope, have the potential to add requirements debt to the system. Changing requirements alters the design of the system, potentially complicating the implementation of existing components and interfaces. This form of requirements debt is especially applicable to adaptive/agile lifecycles which typically start with lower levels of requirements definition. Without a plan for implementation, repeated changes to scope and system intent can significantly delay system development and reduce overall system performance.

Requirements debt is interest-bearing. The developer must adapt to requirements debt, impacting their ability to design and implement the system. Requirements debt can also impact the verification of the system as poorly understood requirements will not be verifiable.

Artifact(s):

- System requirements specifications.

Compromises:

- Not including sufficient stakeholders
- Not reviewing and analyzing the requirements
- Minimizing requirements reviews
- Changing or adding requirements.

Impacts:

- Incorrect or unverifiable designs and implementations
- Rework due to misunderstanding or failed verification
- Increased design cycles to ensure requirements are satisfied.

Example Nelson Mandela Bay Buses.

The Nelson Mandela Bay Metropolitan Municipality desired to upgrade the bus service in Port Elizabeth, South Africa. In 2009 it purchased 60 new buses as part of a $130,000,000 project to implement a bus-based rapid transit system. The buses were used to support the 2010 World Cup. However, following this event, the buses were discontinued from use until significant issues with the buses and transit system could be resolved [34].

The issues with the buses have been traced to poor requirements. The streets contain central islands which were intended to be used as bus passenger drop off locations. However, the specification for the buses did not identify the need to drop passengers off both on the side of the road and on the central island. As a result, the buses only have doors on one side, meaning that the central island infrastructure is unused. Other requirement issues included incorrectly specifying the required bus width and therefore the buses acquired were too big for the bus lanes on the roads. These issues combined to prevent the buses from being used [34].

The poor requirements management process resulted in requirements that did not produce a system that met user needs. Within the context of the transportation system, the bus requirements were not correct: doors were needed on both sides of the bus and the bus width was not consistent with the width of the bus lanes. This requirement debt delayed the release of the bus system, with the buses sitting idle, while the developer attempted to fix the problems. In this case, several different companies and project managers have been used to try to address the problems, compounding the technical debt interest observed from poor requirements.

Technical Compromise:

- Limited requirements review and attention to detail.

Short-term Benefit:

- Acquired in time to support the World Cup.

Long-term Consequences:

- Interest: Additional maintenance for unused buses, alterations to the transit system due to the mismatched designs of the buses and infrastructure
- Fees: Inability of the users to use the bus at all stops.

5.9 Test Debt

Testing is the last step before a system (or component) is deployed and it is usually subject to schedule pressure based on the accumulated delays of the previous system phases. Failures in test cause their own delays, as rework is required to fix the issues. Therefore, test is a process where it is easy to create technical debt: do we really need to run all these tests or is it good enough as is? Test debt occurs when technical compromises are made during the preparation for and execution of system and component test events [35]. It includes concessions made in the creation of test plans and procedures, selection of test steps, and in the selection of the components under test themselves [16, 17].

Example scenarios where test debt can occur in systems engineering include:

Testing at the Wrong Level Verification tests are often completed at the lowest level possible to show that the component functions as designed. However, that does not remove the need to perform integrated tests, or to ensure that each component functions properly when integrated into the system. The failure to perform end-to-end tests and to rely too heavily on component-level testing can result in the system failing to operate as desired in all conditions.

Insufficient Tests With the desire to release systems as quickly as possible, developers may cut back on testing, focusing only on testing nominal paths through the system instead of examining system behavior in off-nominal conditions. Off-nominal tests will reveal system behavior in various real-world scenarios and lacking these tests can result in longer times to diagnose and correct discrepancies identified in operations.

Minimizing Regression Tests Any update or upgrade to the system, whether performed in hardware or software, changes the system baseline. A suite of regression tests needs to be performed to ensure that there are no unintended consequences of the baseline change. Installing new components without performing regression tests can result in system failures affecting the system performance.

Insufficient Component Maturity Testing between system components is critical to ensure that they interoperate properly. However, if a system component is not sufficiently mature, then the test may give false impressions of overall success and may drive the acceptance of workarounds. It is critical that each component involved in a test has demonstrated proper maturity going into the test.

Poor Test Documentation Test documentation is critical both in developing repeatable tests and in capturing the behavior of the system under test. Test procedures without sufficient clarity may result in a tester not understanding how to execute the test and either increasing the test duration or testing the wrong aspects of the system, requiring a retest. If the test results are not properly captured, then the development team may not know how to correct the deficiencies revealed in the test, increasing rework time.

Test debt is interest-bearing as insufficient testing may increase the development time of future parts of the system. Insufficient early-stage testing can lead to longer integration times and fixes at the system level. Test debt is also fee-bearing since the system may be deployed without fully characterizing its behavior in its operational environment. The system may be more complicated to use or underperform, resulting in fees to the system user.

Artifact(s):

- Test plans and procedures
- Test events
- Test reports.

Compromises:

- Incomplete testing at the component and/or system level
- Insufficient or unclear test plans and procedures
- Incomplete test reports.

Impacts:

- Lack of thorough testing leads to problem discovery later in the system lifecycle, increasing difficulty of fixes (interest) and increasing impact to users (fees).

Example Purchasing F-35s before the Completion of Operational Testing.

The F-35 Joint Strike Fighter is an advanced stealth fighter developed by the U.S. DoD and is also the DoD's most expensive weapon system program. While an extremely capable aircraft, its development has been beset by delays of over ten years and $165 billion in cost growth. As of 2022, the F-35 program had not demonstrated acceptable levels of performance and reliability to enable the DoD to authorize the program to enter full-rate production. Some of the problems with the program relate to the development of the F-35 simulation environment, which is used to conduct critical tests which would enable the full-rate production decision. The delays in the simulation environment also delay the completion of operational test and evaluation. Even with these delays, the DoD is still acquiring aircraft at high rates. Acquiring these aircraft prior to completing testing may result in increased work and costs if the tests reveal deficiencies which must be corrected [36].

By deciding to purchase the aircraft to achieve the short-term benefit of fielding them, the DoD is incurring technical debt. Any workarounds required to adapt to failed tests or underperformance represent interest on the technical debt. These workarounds include performing additional tests and associated rework to correct the deficiencies on the already acquired aircraft, if required. The technical compromise is the reduction in the testing performed on the acquired aircraft, and therefore this example represents an instance of test debt.

Technical Compromise:

- Purchasing aircraft prior to completing operational acceptance testing.

Short-term Benefit:

- Fielding aircraft.

Long-term Consequences:

- Interest: Additional testing and resulting rework to correct any problems found during test on the acquired aircraft
- Fees: Untested issues experienced by aircraft pilots that reduce the F-35 performance.

5.10 Summary

In this chapter, we reviewed several types of technical debt that are relevant to systems engineering and associated them with the artifacts that they affect. By associating types of technical debt with specific system artifacts, we are able to thread the needle between creating too many and too few types of technical debt. We have created a list of technical debt types that can be used to further our understanding of how technical debt occurs in the system lifecycle, which we will discuss in the next chapter.

References

1. Kruchten, P., Nord, R. L., & Ozkaya, I. (2012). Technical Debt: From Metaphor to Theory and Practice. *IEEE Software, 29*(6), 18–21.
2. Kruchten, P., Nord, R., & Ozkaya, I. (2019). *Managing Technical Debt: Reducing Friction in Software Development*. Addison-Wesley Professional.

3. Kleinwaks, H. (2023). *Avoiding Technical Bankruptcy in System Development: A Process to Reduce the Risk of Accumulating Technical Debt.* Doctoral Dissertation, Colorado State University, Department of Systems Engineering.
4. Calleam Consulting, Ltd. (n.d.). *Catalogue of Catastrophe.* Calleam Consulting, Ltd. Retrieved April 8, 2025, from https://calleam.com/WTPF/?page_id=3
5. Walden, D. D., Roedler, G. J., Forsberg, K. J., Hamelin, R. D., & Shortell, T. M. (Eds.). (2015). *Systems Engineering Handbook* (4th ed.). John Wiley and Sons.
6. Verdecchia, R., Malavolta, I., & Lago, P. (2018). Architectural Technical Debt Identification: the Research Landscape. *2018 ACM/IEEE International Conference on Technical Debt*, 11–20. https://doi.org/10.1145/3194164.3194176
7. Soliman, M., Avgeriou, P., & Li, Y. (2021). Architectural design decisions that incur technical debt – An industrial case study. *Information and Software Technology, 139.* https://doi.org/10.1016/j.infsof.2021.106669
8. Dooley, R. (2023, January 6*). Southwest's $825 Million Loss Signals Why Companies Shouldn't Put Off Internal Investments.* Forbes. https://www.forbes.com/sites/rogerdooley/2023/01/06/southwest-meltdown-could-cost-825-million/
9. Wikipedia contributors. (2025, March 2). *2022 Southwest Airlines scheduling crisis.* Wikipedia, The Free Encyclopedia. Retrieved April 7, 2025, from https://en.wikipedia.org/wiki/2022_Southwest_Airlines_scheduling_crisis
10. Tom, E., Aurum, A., & Vidgen, R. (2013). An exploration of technical debt. *The Journal of Systems and Software, 86*, 1498–1516. https://doi.org/10.1016/j.jss.2012.12.052
11. Porter, J., & Vincent, J. (2022, June 7). *USB-C will be mandatory for phones sold in the EU 'by autumn 2024'.* The Verge. Retrieved April 7, 2025, from https://www.theverge.com/2022/6/7/23156361/european-union-usb-c-wired-charging-iphone-lightning-ewaste
12. Savov, V. (2011, February 9). *European Union's 'One charger for all' starts sampling, coming this year.* Engadget. Retrieved April 7, 2025, from https://www.engadget.com/2011-02-09-european-unions-one-charger-for-all-starts-sampling.html?guccounter=1
13. Priestley, T. (2016, January 11). *Apple Ditching The Headphone Jack Is Less About Music, More About Royalties.* Forbes. Retrieved April 7, 2025, from https://www.forbes.com/sites/theopriestley/2016/01/11/apple-ditching-the-headphone-jack-is-less-about-music-more-about-royalties/
14. Davidson, J. (2023, September 15). *With phone sales falling, Apple prepares a plan B.* The Australian Financial Review. Retrieved April 7, 2025, from https://www.afr.com/technology/with-phone-sales-falling-apple-prepares-a-plan-b-20230914-p5e4r6#:~:text=much%20$US5%20billion%20($7.8%20billion)%20a%20year,by%20shifting%20to%20the%20vanilla%20USB%2DC%20cable
15. Wikipedia contributors. (2025, March 18). *Version control.* Wikipedia, The Free Encyclopedia. Retrieved April 7, 2025, from https://en.wikipedia.org/wiki/Version_control
16. Li, Z., Avgeriou, P., & Liang, P. (2015). A systematic mapping study on technical debt and its management. *The Journal of Systems and Software, 101*, 193–220. https://doi.org/10.1016/j.jss.2014.12.0-27
17. Rosser, L. A., & Ouzzif, Z. (2021). Technical Debt in Hardware Systems and Elements. *IEEE Aerospace Conference (50100)*, 1–10. https://doi.org/10.1109/AERO50100.2021.9438332
18. Wikipedia contributors. (2025, April 6). *Airbus A380.* Wikipedia, The Free Encyclopedia. Retrieved April 7, 2025, from https://en.wikipedia.org/wiki/Airbus_A380
19. Calleam Consulting, Ltd. (n.d.). *Airbus – A380.* Calleam Consulting, Ltd. Retrieved April 7, 2025, from https://calleam.com/WTPF/?p=4700
20. Storrle, H., & Ciolkowski, M. (2019). Stepping Away from the Lamppost: Domain-Level Technical Debt. *2019 45th Euromicro Conference on Software Engineering and Advanced Applications (SEAA).* 325–332. https://doi.org/10.1109/SEAA.2019.00056
21. Calleam Consulting, Ltd. (n.d.). *Avon Products.* Calleam Consulting, Ltd. Retrieved April 7, 2025, from https://calleam.com/WTPF/?p=6248
22. Henschen, D. (2013, December 16). *Inside Avon's Failed Order-Management Project.* Information Week. Retrieved April 7, 2025, from https://www.informationweek.com/data-management/inside-avon-s-failed-order-management-project

23. Kimberley, O. (2025, March 10). *It's Time For Marketers To Redefine The RFP As The 'Request For People'*. Forbes. Retrieved April 7, 2025, from https://www.forbes.com/sites/quad/2025/03/10/its-time-for-marketers-to-redefine-the-rfp-as-the-request-for-people/
24. Wikipedia contributors. (2025, March 13). *Samsung Galaxy Note 7*. Wikipedia, The Free Encyclopedia. Retrieved April 7, 2025, from https://en.wikipedia.org/wiki/Samsung_Galaxy_Note_7
25. Byford, S. (2017, January 22). *Samsung says bad batteries and rushed manufacturing doomed the Galaxy Note 7*. The Verge. Retrieved April 7, 2025, from https://www.theverge.com/2017/1/22/14353686/samsung-galaxy-note-7-recall-reason-report
26. Vincent, J. (2017, January 20). *Samsung Note 7 investigation will blame 'irregularly sized' batteries and manufacturing flaws, says WSJ*. The Verge. Retrieved April 7, 2025, from https://www.theverge.com/2017/1/20/14334376/samsung-note-7-flaw-batteries-wsj-report
27. Yang, Y., Michel, R., Wade, J., Verma, D., Torngren, M., & Alelyani, T. (2019). *Towards a taxonomy of technical debt for COTS-intensive cyber physical systems*. Procedia Computer Science, 153, 108–117.
28. Bellringer, C. (2015). *An Audit of the Panorama Public Health System*. Victoria, British Columbia: Office of the Auditor General of British Columbia.
29. Calleam Consulting, Ltd. (2020, April). *Panorama*. Calleam Consulting, Ltd. Retrieved May 25, 2025, from https://calleam.com/WTPF/?p=9352
30. Oberg, J. (1999, December 1). *Why the Mars Probe went off course*. IEEE Spectrum. Retrieved April 7, 2025, from https://spectrum.ieee.org/why-the-mars-probe-went-off-course
31. IBM. (2026, January 29). *IBM Engineering Requirements Management DOORS*. IBM. Retrieved February 21, 2026, from https://www.ibm.com/docs/en/engineering-lifecycle-management-suite/doors/9.7.2?topic=engineering-requirements-management-doors-overview
32. Lenarduzzi, V., & Fucci, D. (2019). Towards a Holistic Definition of Requirements Debt. *2019 ACM/IEEE International Symposium on Empirical Software Engineering and Measurement (ESEM)*, 1–5.
33. Pohl, K. (2010). *Requirements Engineering: Fundamentals, Principles, and Techniques* (1st ed.). Springer Publishing Company, Incorporated.
34. Calleam Consulting, Ltd. (n.d.). *Nelson Mandela Bay Metropolitan Municipality*. Calleam Consulting, Ltd. Retrieved May 3, 2025, from https://calleam.com/WTPF/?p=8149
35. Alves, N. S., Ribeiro, L. F., Caires, V., Mendes, T. S., & Spinola, R. O. (2014). Towards an Ontology of Terms on Technical Debt. *2014 Sixth International Workshop on Managing Technical Debt*. 1–7.
36. United States Government Accountability Office. (2022). *F-35 Joint Strike Fighter: Cost Growth and Schedule Delays Continue*. (GAO-22-105943).

Chapter 6
Technical Debt in the Systems Engineering Lifecycle

Abstract Systems engineering projects all move through a defined system lifecycle, traditionally represented as the systems engineering "Vee" diagram. Even with Agile and incremental development models, the same basic stages occur. When investigating technical debt within systems development, it is important to understand both how technical debt is created in each stage of the system's lifecycle and when it is likely to be observed. Specifics of technical debt creation in incremental and adaptive models are presented. An overview of the schedule impact of interest-bearing technical debt is provided. The impact of fee-bearing debt on system acceptance is discussed.

Keywords Technical debt in systems engineering lifecycle · Estimate of schedule impacts

We have spent the last several chapters defining the aspects of technical debt, its causes, and its types. In this chapter, we put these topics together and discuss how technical debt influences the different stages of the systems engineering lifecycle.

When considering technical debt in the course of system development, we must assess the following topics:

- Where in the lifecycle is technical debt created?
- Where in the lifecycle is technical debt observed?
- What aspects of the lifecycle model lend itself to the discovery and management of technical debt?

First, let's return to the results of the survey of professional systems engineers discussed in Sect. 1.4. The results, shown in Fig. 6.1, indicate that technical debt is more likely to be created early in the system lifecycle and its impacts are more likely to be observed late in the lifecycle. Of particular note is the very high level of technical debt observed in operations, which is indicative of technical fees.

We start this chapter by discussing technical debt within the stages of the systems engineering lifecycle. We then look at additional considerations for technical debt management in iterative and adaptive lifecycles. Finally, we conclude the chapter

H. Kleinwaks, *Managing Technical Debt in Systems Engineering*,
https://doi.org/10.1007/978-3-032-30360-8_6

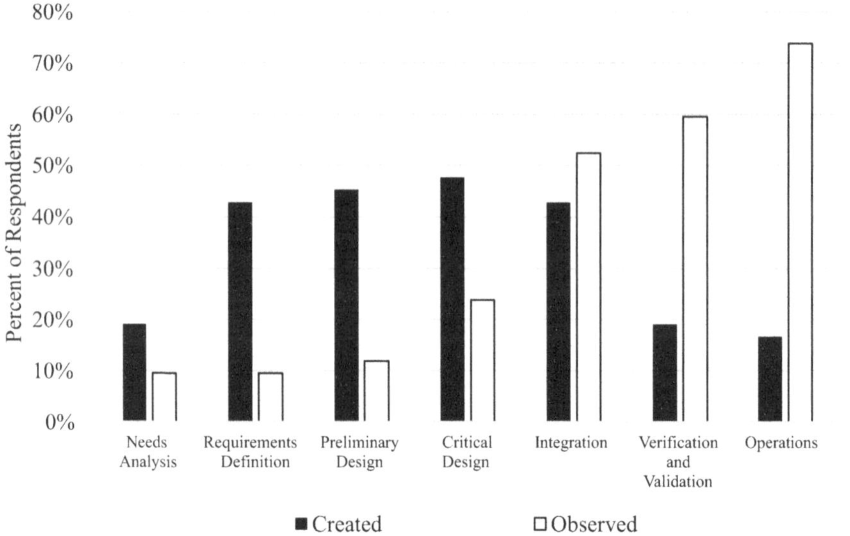

Fig. 6.1 Technical debt in the system lifecycle [1]

with a discussion of how technical debt impacts both project schedules and system acceptance.

6.1 Technical Debt in the Stages of the Systems Engineering Lifecycle

Given that a technical debt item must be associated with a system artifact, we can begin to understand where different types of technical debt will appear in a system lifecycle based on the artifacts generated in each stage. In the following sections, we'll discuss the artifacts created and consumed in each of the lifecycle stages and how this informs the types of technical debt that are associated with each stage. Fig. 6.2 depicts the types of technical debt created (left side, light gray) and observed (right side, dark gray) in each stage of the system lifecycle.

Interest-bearing technical debt affects the system developer, but not the system user. Therefore, we would expect to see a large number of interest-bearing technical debt types affect the system up through verification and validation. Indeed, 7 of the 9 technical debt types are likely to be observed in the verification and validation phase. Fee-bearing technical debt affects the operations phase, most specifically domain debt. Domain debt accounts for the majority of the issues seen during operations when we think of issues with validation. The system not performing as intended in its application domain is domain debt, and a lot of issues can be mapped to domain

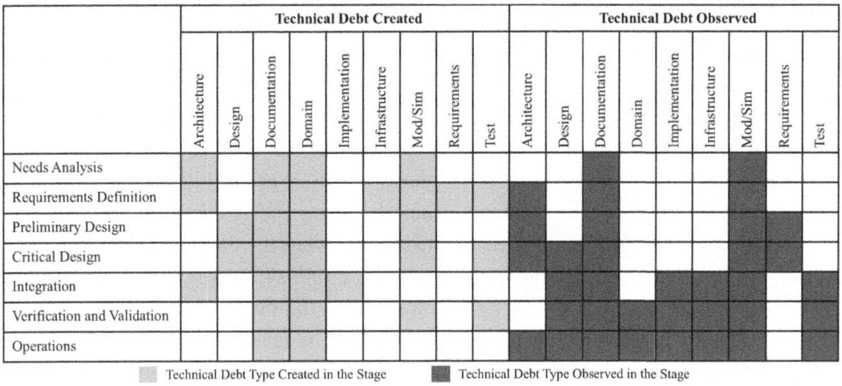

Fig. 6.2 Technical debt creation and observation within the system lifecycle, adapted from [2]

debt. However, many other types of technical debt are fee-bearing, which accounts for their observation during operations.

With this basic understanding of the flow of technical debt, let's look a little more closely at each stage. We will examine the typical artifacts created and used and identify how technical debt can be incurred during the phase.

6.1.1 Needs Analysis

In the needs analysis stage, the initial concepts of operations and architecture level trades are completed such that detailed requirements for a system which will meet the needs of the stakeholders can be created. The needs analysis stage is the first stage of the system lifecycle process, and therefore it does not use any preexisting artifacts. The artifacts created during this stage include the description of capabilities and the statement of operational effectiveness, which describe the goals and expected performance of the system. The initial system architecture concepts and architecture level models and simulations are also created and socialized with the stakeholders to achieve concurrence on the concepts.

Domain debt can easily be created during this stage if sufficient agreement between the stakeholders and the developer is not captured in these foundational documents. Any misunderstandings about the application domain at this early stage will grow into significant problems later in the project. These misunderstandings may not reveal themselves until the validation stage, when it is too late to easily correct them. Architecture debt can be created in this stage due to a reliance on existing products and services that do not actually meet the stakeholder needs. If initial models and simulations are of insufficient fidelity to demonstrate the utility (or lack thereof) of the architecture concepts, then modeling and simulation debt can lead to the selection of an impractical architecture.

The needs analysis stage is one that is often cut short due to various types of pressure. Key stakeholders, including those funding the project, may be eager to get the project moving, putting cost and schedule pressure on the developer. Additionally, political pressure can occur in this stage, where the key stakeholder believes they know the solution better than the system users and ignores initial user feedback. This key stakeholder, often someone responsible for coming up with the idea or leading the design efforts, drives the artifacts created during the needs analysis phase to represent what they think the user needs instead of what the user actually needs. An example of this effect is the Ford Edsel. The Edsel's major failure was due, in part, to a lack of proper market research and testing with potential buyers, resulting in designs that discouraged buyers [3]. Instead of conducting research to develop a car that resonated with buyers, Ford thought they knew better and put the political pressure on the development team to implement what turned out to be a flawed design.

Artifacts used [4]

- None

Artifacts created [4]

- Description of capabilities
- Statement of operational effectiveness
- Initial system architecture concepts
- Architecture models and simulations.

Types of technical debt created

- Architecture debt: the initial system architecture can leverage components that are insufficient, such as COTS or proprietary components, artificially constraining the system
- Domain debt: if the stakeholder needs are not well understood then the description of capabilities and the statement of operational effectiveness, and all of their derived products, will be inaccurate
- Modeling and simulation debt: models of insufficient fidelity can produce inaccurate architectures or unachievable operational effectiveness goals.

Causes of technical debt

- Cost and schedule pressure: the desire to start the project to release it faster and cheaper can cause shortcuts in the needs analysis process, reducing the quality of the gathered information
- Political pressure: a key stakeholder insists that their idea is the way to build the system and that a true understanding of the needs of the users is not required, resulting in domain debt.

6.1.2 Requirements Definition

The requirements definition stage leverages the artifacts generated during the needs analysis stage to create an implementable set of requirements. There are many iterations of requirements definition, as the top-level requirements are generated in collaboration with stakeholders and then the requirements are decomposed for each of the subsystems and components within the system. Once complete, the system (and subsystem) performance requirements and specifications are agreed to between the developer and the stakeholders, forming the basis against which the system will be verified. The methods for verification and the verification events are recorded in the requirements verification traceability matrix (RVTM).

The most obvious form of technical debt created during the requirements definition stage is requirements debt. Poorly formed requirements significantly affect the system developer in multiple ways. Ambiguous requirements may result in a misinterpretation of the intent of the requirement by the developer which then requires additional work to correct. The developer may implement a requirement as written, but if the requirement was not well stated, then the performance may not be what the stakeholder expects. For example, consider the following requirement: "The car shall be capable of parking without driver intervention." This requirement implies an automatic parking feature; however, it does not specify the environmental conditions or type of parking spots that must be supported. The developer can implement a feature where the car can park in a straight-ahead spot in clear daylight only and still meet the letter of the requirement but not the intent. This difference between the intended behavior and the written requirement is the source of the majority of requirements debt. The requirements debt appears in the performance requirements and functional specifications created during this stage.

Other types of technical debt may also be created during the requirements definition stage. Infrastructure debt arises when requirements do not properly specify the supporting infrastructure required to execute the project. Test debt appears in the RVTM when the verification methods are not adequate for the requirements being verified.

Driving factors for the accrual of technical debt in the requirements definition phase include cost and schedule pressure. The system developer and the stakeholders are both eager to get the design of the project started and therefore may not properly capture requirements. Schedule and cost pressure and technical trades manifest in the requirements statements themselves. To use existing systems and limit non-recurring engineering, performance requirements may be reduced from desired values to ensure a system that can be delivered on time and budget, a technical compromise that impacts stakeholders in the long-term.

Artifacts used [4]

- Description of capabilities
- Statement of operational effectiveness
- Initial system architecture

- Architecture models and simulations.

Artifacts created [4]

- System architecture
- System and subsystem performance requirements
- System and subsystem functional specifications
- Conceptual system models and simulations
- RVTM

Types of technical debt created

- Architecture debt: requirements force the system architecture to leverage insufficient COTS products or proprietary structures
- Domain debt: improper reflection of the intended use or performance of the system in performance requirements and functional specifications, including missing requirements for system quality
- Infrastructure debt: insufficient specification of required supporting infrastructure for development, test, and execution in performance requirements and functional specifications
- Requirements debt: poorly formatted requirements in specification documents
- Modeling and simulation debt: insufficient fidelity of modeling and simulation leads to requirements that conflict with each other or are not achievable
- Test debt: verification methods and events specified in the RVTM are not sufficient to verify the system, meaning that proper test plans and equipment will not be developed prior to the start of testing.

Causes of technical debt

- Cost and schedule pressure: stakeholder desire to start design stages forces the developer to complete the requirements definition stage without full requirements verification and analysis
- Technical trades: constraining the cost and schedule to develop the system forces compromises on performance requirements, resulting in a less-performant system than originally intended.

6.1.3 Preliminary Design

In the preliminary design stage, performance and functional specifications are built out into the initial system design. The system developer develops the architecture concepts into a system that provides the required performance while adhering to known system constraints. The developer also must ensure that the design is executable within the allocated cost and schedule. This stage starts the process of creating the system itself from the architecture and is one of the first stages to observe technical debt created in earlier stages. In this stage requirements debt starts to be observed if the initial system design cannot satisfy inconsistent requirements.

Ambiguous requirements will force additional discussion between the system developer and the stakeholders. Architecture debt appears when the developer must 'design around' architecture concepts that were provided in the system architecture.

During the preliminary design stage, the system developer creates drafts of the system documentation and the preliminary design specifications. Architecture models and simulations evolve into preliminary models and simulations used to analyze system performance as the design evolves.

Design debt starts to appear during the preliminary design stage. If the system developer attempts to 'get ahead' by starting work prior to finishing the design, they are creating design debt. With drafts of system documentation comes documentation debt—insufficient or incomplete specification of interfaces and design parameters. Since the documents are typically in draft form, the documentation debt is not as severe as it would be at critical design review, but it does get introduced into the system at this point. Domain debt can also be created in the preliminary design stage if the application domain is not well-understood, resulting in designs that are incompatible with user needs. Finally, modeling and simulation debt can be created if the models and simulations used to predict system performance are not matured along with the system design and if the fidelity is not increased to satisfactory levels.

Technical trades begin to have a significant impact on the system developer during the preliminary design stage. The developer must contend with system constraints that prohibit the preferred design from being realized and therefore must execute technical trades. These trades may reduce performance in one area in order to achieve the requirement in another area. Similar to previous stages, stakeholders may push for rapid closure to the preliminary design stage, putting cost and schedule pressure on the developer. The developer must also contend with the cost and schedule implications of their design. If the design is not within the budget or schedule, then they will need to execute technical compromises to reduce the performance until the design comes back within these parameters. Political pressure is a factor in this stage, as a key stakeholder may insist on a specific design, even if it forces compromises in other areas.

Artifacts used [4]

- System and subsystem performance requirements
- System and subsystem functional specifications
- System architecture
- Architecture models and simulations.

Artifacts created [4]

- System documentation
- Preliminary design specifications
- Preliminary system models and simulations.

Types of technical debt created

- Design debt: inclusion of restrictive interfaces in the design and starting implementation prior to finishing design can introduce design debt

- Documentation debt: poor system documentation of the preliminary design affects critical design and future system development
- Domain debt: misunderstandings of the application domain produce technical trades that do not benefit the user or meet their needs
- Modeling and simulation debt: insufficient fidelity of modeling and simulation leads to designs that will underperform expectations when actually implemented.

Causes of technical debt

- Cost and schedule pressure: the system developer must design a system that is executable within the program cost and budget, forcing compromises between development schedule and system performance
- Technical trades: design choices force technical trades as the limitations of the design are realized. The system developer must make tough decisions about where to meet requirements and where other alternatives must be accepted
- Political pressure: key stakeholders insist that they know the best way to design the system, even if it forces technical compromises in other areas.

6.1.4 Critical Design

The critical design stage refines the preliminary design into the final design and matures it to the point that production can begin. Critical design is where the interface control documents are finalized, detailed models and simulations are used to predict final performance, and technical trades are closed. This stage leverages the artifacts created during the preliminary design stage and the requirements definition stage to create the final system design. Most of these artifacts, such as the design specifications and models and simulations are refined, allowing an opportunity for the system developer to repay technical debt that was created during the preliminary design stage. Note that the developer may observe the effects of technical debt accrued during preliminary design during critical design as critical design efforts can be complicated by existing technical debt, especially in the preliminary documentation and design artifacts.

According to the survey of practicing systems engineers, critical design is the stage in the system lifecycle where technical debt is most likely to be created [1]. This occurs because the critical design stage finalizes the artifacts, which locks in any technical debt which will then be observed during implementation, verification and validation, and operations. Interface control documents must be carefully evaluated for documentation debt, as unclear interface specifications can lead to failures during implementation and test. Test plans, which rely heavily on the RVTM, are finalized. Any test or requirements debt present in the RVTM will be carried forward into the test plans, unless the test developer is properly managing and mitigating those technical debt items. Similarly to preliminary design, insufficient modeling and simulation can introduce technical debt where the system is predicted to have greater performance than it actually will, resulting in underperformance and technical fees.

Technical trades are the driving source of technical debt in the critical design stage, as the developer needs to complete all trade analyses to finish the design. Cost and schedule pressures will appear if the design cannot satisfy all of the performance specifications on time and on budget, resulting in technical compromises. While these pressures are similar to those that occur during preliminary design, the system developer does not have a following design stage in which technical compromises can be resolved, increasing the importance of minimizing the technical debt generated during critical design.

Artifacts used [4]

- System and subsystem performance requirements
- System and subsystem functional specifications
- Preliminary design specifications
- Preliminary system models and simulations
- RVTM

Artifacts created [4]

- System documentation
- Detailed design specifications
- Detailed system models and simulations
- Interface control documents
- Test and evaluation plan

Types of technical debt created

- Design debt: technical trades made to realize schedule or cost benefit may impact the overall design of the system, reducing performance
- Documentation debt: poor system documentation of the final design affects production. Incomplete or insufficient ICDs can lead to incompatible components and test failures
- Domain debt: continued misunderstanding of the application domain results in a design that will not pass validation
- Modeling and simulation debt: insufficient modeling and simulation can result in key performance parameters and figures of merit that are not indicative of true system behavior
- Test debt: incomplete or insufficient test plans will fail to verify the system, resulting in lower quality systems and technical fees.

Causes of technical debt

- Cost and schedule pressure: realities of production time and budget may force technical compromises in the final design.
- Technical trades: completion of technical trades during critical design may force trades between performance across different figures of merit
- Political pressure: key stakeholders insist that they know the best way to design the system, even if it forces technical compromises in other areas.

6.1.5 Integration

In the integration stage, the subsystems and components are built and integrated into the final system. A successful integration stage requires that subsystems and components are compliant with their ICDs and that the developer of each side of the interface interpreted the ICDs in the same way. The integration stage uses ICDs and detailed design specifications to manufacture components, observing any documentation or design debt with these artifacts. These types of debt can complicate both production and integration of the subsystems and components. Infrastructure debt introduced during the requirements development phase is observed during implementation as the lack of proper supporting infrastructure can delay the ability to integrate the system. As an example, consider a construction company that is building a house. The ability to deliver supplies as they are needed, and to have enough supplies, is part of the supporting infrastructure. If the construction company did not spend enough time planning out the logistics, then they may experience delays during the integration phase while waiting for the correct supplies.

During integration, the developer creates system documentation, such as final training materials and as-built specifications for the system. The integration stage is where the system itself is created, introducing implementation debt. Implementation debt can occur due to technical compromises made during the building of the system, such as the lack of quality controls. Such a lack of controls can result in parts being produced out of compliance with the specification, forcing rework. Poor quality controls can result in systems that break more frequently than intended, increasing the maintenance and reducing the quality of the system, producing technical fees. Architecture debt can be introduced if prototype systems are demonstrated and rushed into production instead of implementing the intended design.

During integration, cost and schedule pressure are the primary causes of technical debt. Stakeholders are eager to see the system developed and may have hard dates set on the schedule for system delivery—resulting in schedule pressure to release a system, even if it requires technical compromises. Budget pressures also exist as extending integration times produces additional costs to the stakeholders.

Artifacts used [4]

- Detailed design specifications
- Interface control documents.

Artifacts created [4]

- System documentation
- Components and integrated system
- System specifications.

Types of technical debt created

- Architecture debt: relying on prototypes instead of integrating the final design produces architecture debt

- Documentation debt: insufficient specifications and documentation can make the system more difficult to use or maintain, reducing system quality
- Domain debt: integration choices that do not reflect the needs of the user reduce the usability and insert technical fees
- Implementation debt: compromises made during system implementation can complicate testing and system usability.

Causes of technical debt

- Cost and schedule pressure: stakeholders have target budgets and delivery dates that may force compromises on the developer to enable them to release an integrated system.

6.1.6 Verification and Validation

During verification and validation, the system is determined to be compliant to its requirements (verification) and suitable for use (validation). Verification executes the verification events defined in the RVTM, including inspecting delivered products, running analyses, producing demonstrations, and executing tests. Any test debt or requirements debt present in the RVTM will be observed during this stage, potentially resulting in the need for additional verification efforts. Most other forms of technical debt are also observed during this phase, as verification and validation put the completed system and its documentation under test. The validation process is where domain debt starts to be observed, as users test the system to determine if it meets their needs.

While many types of technical debt are observed during verification and validation, test debt, documentation debt, and modeling and simulation debt are created. Test debt is created through any compromises made during the execution of the tests themselves, either by not doing the tests or not sufficiently testing a part of the system. In the space domain, the goal of testing is to "test like you fly," which means mimicking actual operating conditions as closely as possible during tests. In some cases, a "test like you fly" exception is requested for specific circumstances that cannot be replicated on the ground, such as doing some spacecraft tests in a 1-g gravity field instead of a zero-g gravity field. Too many "test like you fly" exceptions are an indicator that there is technical debt in the system—it is not being tested in a relevant environment. Documentation debt occurs if test results are not properly documented, including failures, to enable a thorough analysis of performance. Modeling and simulation debt occurs if the previous models and simulations are not updated to account for observed performance impacts resulting in inaccurate future performance predictions.

During verification and validation, cost and schedule pressure from stakeholders can force the system developer to compromise on test execution. Tests take time and money to run, and the ability to rerun tests or exhaustively test all failure cases may be limited by the stakeholders. Political pressure comes from key stakeholders who push for test like you fly exceptions. They may be pushing for these exceptions based on past experience or because they are concerned with what the test results may reveal. In either case, limiting the tests introduces the risk that the system may not perform as expected in its operational environment.

Artifacts used [4]

- RVTM
- Test and evaluation plan
- Components and integrated system
- Interface control documents
- Detailed system models and simulations.

Artifacts created [4]

- System documentation
- Test reports
- Verification and validation reports.

Types of technical debt created

- Documentation debt: insufficient documentation of test results complicates failure analysis and repeatability and restricts the traceability of performance of specific production items
- Modeling and simulation debt: failure to update models and simulations based on observed test performance limits the ability to accurately predict operational performance
- Test debt: any minimization of tests risks failing to uncover issues, potentially increasing the maintenance and decreasing the system performance and quality.

Causes of technical debt

- Cost and schedule pressure: shortening the verification and validation cycle leads to an earlier product release, saving time and money but resulting in potential failures in operations
- Political pressure: key stakeholders may press to minimize testing because they 'know it will work' or because tests may reveal poor performance.

6.1.7 Operations

Operations is the stage where the system is used in its application domain. Users of the system leverage the system and its documentation to receive the promised performance. As such, the operations phase does not directly create technical debt, but it does observe multiple forms of technical debt. Most prevalent during operations is domain debt—if the technical compromises resulted in a misunderstanding of the application domain, then the system will not perform as expected, resulting in substantial technical fees. Reduced system quality, which also produces technical fees, can result from implementation or design debt. Unexpected failures may be a product of test debt.

The users may generate feedback to the system developer through formal reports or through informal methods such as product ratings. This feedback is a valuable source of identifying technical fees, which the developer can then trace back to the original technical compromise and cause of the technical debt. Future releases may be able to repay the principal that led to the fees, but the technical fees cannot be removed without an additional release. Therefore, the system developer needs to pay attention to fee-causing debt, and the associated artifacts, throughout the system development process.

Artifacts used [4]

- Integrated system
- System documentation.

Artifacts created [4]

- User feedback reports.

6.2 Technical Debt in Incremental Models

In the previous section we stepped through the stages of a sequential lifecycle model. An incremental lifecycle model uses the same stages, but instead of developing and delivering the entire system at once, each increment develops and delivers a part of the final system. The first increment is especially important in new systems, as the end user must decide if the first increment delivers enough improvement in performance over existing systems to be worth the investment required to produce the new system [5]. While incremental models are subject to similar types of technical debt in similar stages as a sequential model, the incremental model of development introduces the potential for additional technical debt.

This additional technical debt is based on the nature of the incremental model—future increments depend on the work completed in previous increments. Therefore, if an increment makes technical compromises, the future increments will need to work around those compromises. Technical compromises in early increments must balance inputs from the various stakeholders to not only show initial value, but also to prevent the sabotage of future capabilities. Therefore, selection of lower risk and higher technology readiness level[1] components in early increments can be beneficial [5], as long as it does not result in architectural lock-in, producing architecture debt. Future increments must adapt their architecture and design to account for shortcomings in

[1] Technology readiness level (TRL) is a scale developed by NASA to indicate the readiness of a system or component to operate in its intended environment. Low-TRL components (1–3) show completion of analytical studies which validate the proof of concept. Mid-TRL components (4–7) have been prototyped in the laboratory and operational environment. High-TRL components (8–9) are flight-qualified and have completed actual mission operations [6].

previous increments, creating technical debt interest in the form of extended design and development timelines to work around the compromise and fees in the form of overall underperformance of the system.

Architecture debt in incremental development can manifest as many of the exemplars cited in Sect. 5.1. Increments that develop an MVP are prone to having that MVP become incorporated into the final system. Similarly, workarounds developed in one increment may persist into future increments. The system is then susceptible to architectural lock-in, where early increments produce capability that is hard to modify, resulting in the continued use of a less-than-optimal architecture.

Therefore, when using incremental development models, it is critical for the system developer to pay attention to the potential architecture debt that occurs as decisions are made with each increment. Stakeholders will want to see early increments succeed to justify the product development costs. System developers need to resist the pressure to make compromises to achieve early success which have significant negative impacts on the ability to complete future increments and the final system. The system developer must properly allocate requirements across the increments and hold to those requirements to ensure the final system functions as intended.

Let's look at how technical debt can appear in incremental development models. During development of complex systems consisting of multiple components, the primary system developer will derive requirements for each of its subsystems. Subsystems are developed and tested independently to verify those derived requirements. Following these tests, the integrated system is assembled and tested. This separation of development and testing is a form of incremental development.

This type of incremental development is especially susceptible to test debt. If subsystems are tested independently, they may meet their individual requirements. However, when integrated into the system, there may be additional problems discovered. This lack of end-to-end testing was one of the contributing factors to the initial failure of the Hubble Space Telescope.

The Hubble Space Telescope launched in 1990. Two months after its launch, NASA announced that both of the onboard cameras were subject to spherical aberration, which resulted in blurry images [7]. To repair the issue, NASA launched a Hubble servicing mission in December 1993, costing hundreds of millions of dollars [8]. What went wrong?

The primary mirror was manufactured to the wrong tolerance, resulting in an error of 1.3 mm on a 2.4-m mirror. This small offset was enough to make the images taken by the telescope unusable [7]. While this mistake was an error in implementation, the developers of the optical telescope assembly (OTA) also did not plan to conduct any end-to-end tests of the assembly, which would have caught the error, because "an end-to-end test of the OTA would have been very expensive to perform at the level of accuracy specified for the telescope. The test would have cost on the order of what the OTA itself cost" [9]. Instead, a smaller test was performed over a 0.3 m diameter to check the focus. This test was insufficient to verify the focus of the entire mirror [9]. This is a prime example of test debt: because of cost pressures, a technical compromise was made to not test the integrated system, meaning that there was an increased likelihood of the system not performing as planned when all increments were completed and assembled. These consequences were realized and the cost to repair the defect was significant.

In this case, the increment (the OTA assembly) successfully completed the tests it was required to do; however, those tests were insufficient, especially when looking at the performance in the context of the integrated system. Operational conditions were not used for the test, which would have been complicated to perform due to the mirror size. The technical compromise made resulted in unverified increment behavior making its way through the rest of system integration and into the final product, creating disastrous results.

6.3 Technical Debt in Adaptive Models

Adaptive models share similar technical debt potential as sequential and incremental models—they still have the same stages as a sequential model, although those stages may be shorter and less formal. Adaptive models using sprints, such as Scrum, develop a portion of the product each sprint and therefore are subject to the same architecture debt concerns as incremental models. Adaptive models add another potential source for technical debt through their interactive nature.

Adaptive models value frequent interactions with users to get feedback on the system development and to minimize the risks of domain debt. These interactions ensure that the system developer understands the intended use cases and application domain and therefore makes design decisions to deliver a system that is suitable for the user. However, feedback from users can complicate system development, as it may add scope and delay implementation of critical development items which have low visible value to the user.

This pressure from the users is a form of political pressure. While well-intentioned, the pressure can result in the system developer making compromises with respect to aspects of the system that do not appear to add value to the user. For example, quality control processes may be skipped in order to get the system delivered to the user—the user sees the value in the delivered product but not necessarily in the increase in quality control, even though these processes would reduce maintainability and reliability issues down the road. User-driven suggestions may increase scope and cause conflicts with the project schedule and budget, which can force other technical compromises.

Therefore, when using adaptive methods, the system developer must adhere to rigorous change control processes when receiving user feedback. They need to ensure that the recommendations are within overall project scope and that they are implemented in a manner that does not result in technical compromises and the introduction of technical debt.

To understand how technical debt can appear in adaptive methods of systems engineering, let's consider the seven methods for applying the Agile Manifesto to systems engineering provided by Douglass [10].

1. **Work Incrementally:** incremental development approaches enable the development of separate parts of the system that can be independently verified with frequent user feedback. The introduction of technical debt in incremental development is discussed in Sect. 6.2.
2. **Plan Dynamically:** plan for an iteration and replan as necessary based on the realities of the delivered capabilities in each iteration. The iterative planning approach tends to favor solving the largest current problems without necessarily considering the longer-term consequences of the decisions, thereby creating technical debt.
3. **Actively Reduce Project Risk:** introduce spikes to mitigate risks and then replan to address issues. Replanning can introduce technical debt if the effect on the final system is not properly considered. Therefore, this method should be augmented to include technical debt mitigation in the project spikes in addition to risk mitigation. Section 7.2 provides a more detailed comparison of technical debt management and risk management.
4. **Verify Constantly:** frequent unit, integration, and system level verification is necessary to produce a functioning system. This verification serves to reduce technical debt by catching test failures early.
5. **Integrate Continuously:** frequent integration and verification of interfaces to ensure that the various components work together. This method has the potential to increase technical debt if workarounds (instead of actual fixes) are implemented to enable integration.
6. **Validate Frequently:** enable the users to frequently review the product to ensure that it meets their needs. This method reduces technical debt by ensuring that the system properly represents the application domain.
7. **Modeling is Essential:** building digital models helps to reduce technical debt by exploring the solution set prior to actually implementing the system design in hardware; however, the system developer must be careful not to introduce technical debt into the models themselves.

A team from the University of Luxembourg [11] implemented Agile methods in the development of miniature satellites. They attempted to follow Agile frameworks used in software development through the system development process. As such, their work provides an excellent case study that reveals some of the ways that technical debt appears when using adaptive methods in system development. Specifically, they found:

- At times they had to use different parts as 'stand-ins' when the actual parts were delayed. The desire to produce functioning products with each iteration required the use of these 'stand-in' parts. However, they observed that the team had to be aware of the difference in capability from the 'stand-in' part to have fewer surprises later in the development process. The use of 'stand-in' parts is an example of implementation debt and could significantly affect later iterations. Interest is paid to work around the limitations of the stand-in part and then taxes are paid to replace the stand-in with the actual part.

- Documentation was insufficient to enable rapid on-boarding of new team members. The documentation would quickly become out-of-date due to the rapid development pace and the lack of training materials for new personnel slowed development, instances of documentation debt.
- The development process experienced unplanned events related to the establishment of employee work environments, such as supply chain delays and equipment issues related to the computer and network systems. This infrastructure debt forced work stoppages and led the team to develop a simplified work environment for each team member to mitigate the delays.

6.4 Estimating the Schedule Impact of Interest-Bearing Technical Debt

We discussed the concept of compounding technical debt in Sect. 4.6. Once present in the system, interest-bearing technical debt can expand, affecting multiple artifacts and slowing down system development. The field of project management focuses on delivering systems on time and on budget. Project managers manage the delivery timeline through project schedules which link the various tasks together into a sequence and identify the dependencies between the tasks. Even the best schedule is subject to uncertainty, which project managers handle by adding margin to the duration of one or multiple tasks [12], which can help to mitigate delays that occur with the task execution.

Project schedules are typically shown as Gantt charts where each task is represented on a row and the start and stop times are indicated by the size of the colored bar, as shown in Fig. 6.3. By estimating the duration of each task and the sequencing of the tasks, project schedulers can produce an estimate of when the project will complete. By analyzing the schedule, the tasks which drive the schedule, called the critical path, can be derived. In Fig. 6.3, it is clear that task H cannot start until tasks E, F, and G are complete. The critical path is A- > D- > G- > H- > I- > J as this is the sequence of events that, if a task end date changes, would force the project end date to change. Task F is not on the critical path. Even if Task F slips by two time periods, task H cannot start until task G completes.

As useful as a simple Gantt chart can be, it often relies on several incorrect assumptions. First, it assumes that each task duration is known and that the task will complete within that duration. Second, it assumes that each task completes perfectly, enabling follow-on tasks to execute without problems. The first assumption is accounted for by performing schedule risk assessments (SRA) where a probability distribution is assigned to the duration of each task and then a Monte Carlo analysis is performed. By assessing the outputs of the Monte Carlo runs, end dates can be determined that fall within confidence intervals, enabling the system developer to make claims such as "75% of the time we will finish by March 1 and 95% of the time we will finish by May 15". The system developer can also use the results to

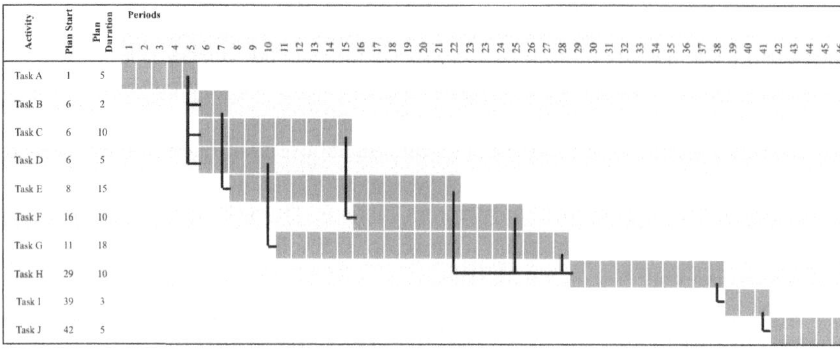

Fig. 6.3 Example Gantt chart schedule

examine changes to the critical path and to identify which tasks are at the highest risk of delaying the project.

While SRAs provide schedule insights for different lengths of tasks, each task is assessed independently. In reality, pressure to complete a task on time may force the developer to make technical compromises which lower the task quality, thereby introducing technical debt into the system. The quality with which one task is completed directly affects the duration and quality of its follow-on tasks, as we will see in the following example.

Technical debt can be included in SRAs by adjusting the duration of tasks based on technical debt incurred from predecessor tasks. The increase in duration is related to the amount of technical debt—the more compromises made in the predecessor task, the more likely the successor task is delayed. Consider the notional project shown in Fig. 6.4. This project consists of ten tasks, and has three separate, parallel development paths that each begin after the completion of the first task (Task A). Below each task are the triangular distribution parameters for the task, where L is the low estimate, M is the medium value, and H is the high value, all of which are specified in days. This project consists solely of finish-to-start dependencies, meaning that the successor tasks cannot complete until the predecessor tasks have completely finished.

Using this project, schedule risk analyses can be run considering three different scenarios: the baseline project completing as intended, Task C introducing technical debt and therefore increasing the duration of Task F, and Task C introducing technical debt which compounds and therefore affects all successors (Tasks F, H, I, and J). The results of these analyses are listed in Table 6.1, and a full treatment of the process is given in Appendix A. We can see that, in this project, making a technical compromise on one task can increase the average project duration by over 25%!

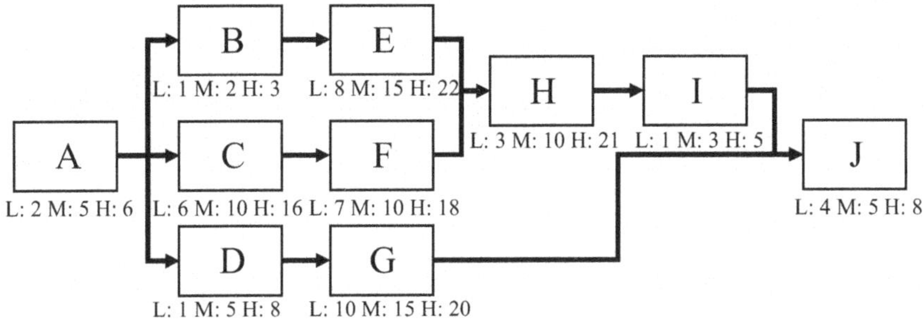

Fig. 6.4 Notional project

Table 6.1 Project duration estimates due to technical debt

	Average Duration	% Increase in Duration	90th Percentile Duration	% Increase in 90th Percentile Duration
Baseline	47 days	N/A	53 days	N/A
Technical Debt from Task C affects Task F only	52 days	11%	60 days	13%
Technical Debt from Task C compounds and affects multiple successors	60 days	28%	71 days	34%

6.5 Impact of Fees on System Acceptance

The presence of technical fees can have a large impact on the acceptance of the system by its users. After all, fees impact the user, not the developer. A developer can create a product that meets all the requirements. If the user does not accept the product, then it will not sell. Why wouldn't a user accept the product? One reason may be the presence of technical fees.

Technical fees are paid by the user each time they use the system. Fees represent reductions in expected system performance, either in scope or quality. If fees accumulate within the system, the user may choose to not use the system. In the commercial world, this impact would be seen as a reduction in sales. In the systems engineering sense, the product would fail validation.

Within the U.S. DoD, there are formal processes to verify and validate systems. Developmental Test (DT) is the verification process, while Operational Test (OT) is the validation process. OT subjects the system to real-world conditions with real-world users to determine if the system is effective, maintainable, reliable, and usable [13]. A system that passes its OT is accepted for operational use.

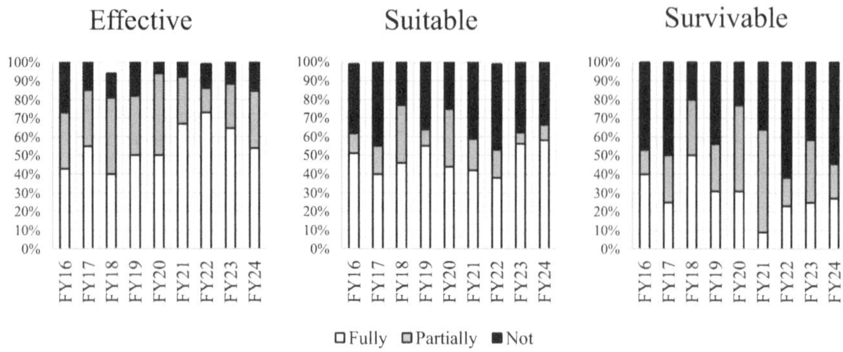

Fig. 6.5 2024 operational performance trends, adapted from [14–16]

The OT evaluation is in the performance dimension: the effectiveness is a measure of the system performance and the system quality is measured by its maintainability, reliability, and usability. Since technical fees are a reduction in the performance dimension experienced by the user, then we can conclude that the presence of technical fees impacts the ability of a system to pass OT and to be accepted.

Figure 6.5 compiles data from Director of Operational Test and Evaluation (DOT&E) reports [14–16] and shows that over a 9-year period, approximately 50% of systems tested were deemed operationally effective, 50% were deemed operationally suitable, and only 30% were deemed survivable. Yet each of these systems completed development and (supposedly) verified their requirements prior to reaching operational test! Technical fees are the underlying reason for the failure of these systems during operational test validation.

To demonstrate the impact of technical fees, I can provide an example from my own career. I worked on a software system that was designed to replace a legacy system. The new system contained updated graphical displays and user interfaces and automated many of the tasks that the previous user had to do manually. We designed this system and its interface without substantial input from the users (domain debt) and produced (what we thought was) a very functional and easy-to-use interface that met the requirements. We leveraged previous work and interfaces that our company was familiar with (design debt) instead of truly understanding the use cases of interest to the system user.

It should be no surprise that when we put the system in front of the users, they were extremely critical of the work. We had complicated some of their tasks, in part because we did not work with them ahead of time to fully understand their required workflows—introducing technical fees as a result of the domain debt. The existing software components we leveraged were largely unused by the users, meaning that the design debt resulted in extra work that we performed and now had to remove—we had to pay taxes. Fortunately for our development team, these interactions occurred early enough in the development cycle where we were able to correct them, and we substantially increased our interactions with the users to better understand their needs. Had this system gone to OT as it was designed, it would have failed user acceptance due to the technical fees, even though it met its requirements.

6.6 Summary

In this chapter, we examined how system artifacts are used to trace technical debt throughout the systems engineering lifecycle. As system lifecycles become more incremental and adaptive, additional sources and types of technical debt manifest themselves. This is the tradeoff with the flexibility of development processes—the more flexible the development process, the more likely there is to be technical debt, as the system design evolves with the changing requirements and component development. We also examined how technical debt impacts project schedules and system acceptance, two critical areas that must be carefully managed to ensure a successful system development.

This chapter concludes the second part of this book. We have now established how technical debt appears within systems engineering projects. In the next part, we will discuss how to manage technical debt within systems engineering to reduce its impacts on the success of system development.

References

1. Kleinwaks, H., Batchelor, A., & Bradley, T. H. (2023). An Empirical Survey on the Prevalence of Technical Debt in Systems Engineering. *INCOSE International Symposium*, 33(1), 1640–1658. https://doi.org/10.1002/iis2.13104
2. Kleinwaks, H. (2023). *Avoiding Technical Bankruptcy in System Development: A Process to Reduce the Risk of Accumulating Technical Debt.* Doctoral Dissertation, Colorado State University, Department of Systems Engineering.
3. Wikipedia contributors. (2025, April 11). *Edsel.* Wikipedia, The Free Encyclopedia. Retrieved April 27, 2025, from https://en.wikipedia.org/w/index.php?title=Edsel&oldid=1284989578
4. Kossiakoff, A., Seymour, S. J., Flanigan, D. A., & Biemer, S. M. (2020). *Systems Engineering: Principles and Practice* (3rd ed.). John Wiley and Sons.
5. Mortlock, R. F. (2020). Studying acquisition strategy formulation of incremental development approaches. *Defense AR Journal*, 27(3), 264–311.
6. NASA Earth Science Technology Offices. (2026). *Technology Readiness Levels.* NASA. Retrieved February 22, 2026, from https://esto.nasa.gov/trl/
7. National Aeronautics and Space Administration. (2025, June 11). *Hubble's Mirror Flaw.* NASA. Retrieved June 11, 2025, from https://science.nasa.gov/mission/hubble/observatory/design/optics/hubbles-mirror-flaw/
8. Gainor, C. (2020). *Not Yet Imagined: A Study of Hubble Space Telescope Operations.* (NASA SP-4237).
9. National Aeronautics and Space Administration. (1990). *The Hubble Space Telescope Optical Systems Failure Report.* (NASA-TM-103443).
10. Douglass, B. P. (2016). *Agile Systems Engineering.* Morgan Kaufmann.
11. Kanavouras, K., Hein, A. M., & Sachidanand, M. (2022). Agile Systems Engineering for sub-CubeSat scale spacecraft. Preprint retrieved from https://arxiv.org/pdf/2210.10653

12. Love, P. E., Ackermann, F., Smith, J., Irani, Z., & Edwards, D. J. (2016). Making Sense of Rework Causation in Offshore Hydrocarbon Projects. *Project Management Journal, 47*(4), 16–28.
13. Air Force Operational Test and Evaluation Center. (n.d.). *Air Force Operational Test and Evaluation Center*. Retrieved July 9, 2022, from https://www.afotec.af.mil/About-Us/Fact-Sheets/Display/Article/872935/air-force-operational-test-and-evaluation-center/
14. O'Toole. Jr., R. D. (2025). *Director, Operational Test and Evaluation FY 2024 Annual Report*. The Office of the Director, Operational Test and Evaluation. Retrieved from https://www.dote.osd.mil/Portals/97/pub/reports/FY2024/other/2024Annual-Report.pdf?ver=AkqD4y1xIhmNndurzRkvqQ%3d%3d
15. O'Toole, Jr., R. (2024). *Director, Operational Test and Evaluation FY 2023 Annual Report*. The Office of the Director, Operational Test and Evaluation. Retrieved from https://www.dote.osd.mil/Portals/97/pub/reports/FY2023/other/2023annual-report.pdf?ver=d7gusiIrcbYmxM0oDkPSFg%3d%3d
16. Guertin, N. H. (2023). *Director, Operational Test and Evaluation FY 2022 Annual Report*. The Office of the Director, Operational Test and Evaluation. Retrieved from https://www.dote.osd.mil/Portals/97/pub/reports/FY2022/other/2022annual_report.pdf?ver=71JCDFcAlC9z_UnuI9BOUQ%3d%3d

Part III
Managing Technical Debt in Systems Engineering

Chapter 7
Managing Technical Debt

Abstract Technical debt management is a five-step process: identification, measurement, prioritization, repayment, and monitoring. Each technical debt item goes through the first four steps, while monitoring is a continuous process applied to the entire system. While technical debt management is similar to risk management, there are several key differences. Risk management attempts to minimize the likelihood or consequence of an uncertain event. Technical debt management seeks to minimize the uncertain consequences of a known event (the technical compromise). To manage technical debt, it is first necessary to develop a technical debt register to properly track different technical debt items in the system. Since technical debt is introduced to the system through unintentional means, methods and techniques are provided to reduce the occurrence of unintentional technical debt.

Keywords Technical debt management cycle · Technical debt management versus risk management · Technical debt register · Unintentional technical debt

With this chapter, we move from discussions about what technical debt is to how to manage technical debt within systems engineering projects. We've already discussed how systems will face pressure from stakeholders and how that pressure will lead to technical debt. There are no processes that we can provide that will remove that pressure or that will prevent technical debt. And remember, not all technical debt is bad. Therefore, the goal of a technical debt management process is not to eliminate technical debt. Instead, by managing our technical debt, we can leverage it for the benefit of the project.

Let's review the technical debt quadrant that we first introduced in Chap. 2. In Fig. 7.1, we've added two new arrows to the quadrant: preparation and programmatic controls. These arrows represent the fundamental components of a technical debt management system. Preparation ensures that the team is ready to find and discuss technical debt items throughout the system lifecycle, while programmatic controls ensure that system events and artifacts account for identified technical debt items. Following a technical debt management process can move technical debt items to the upper right of the quadrant.

H. Kleinwaks, *Managing Technical Debt in Systems Engineering*,
https://doi.org/10.1007/978-3-032-30360-8_7

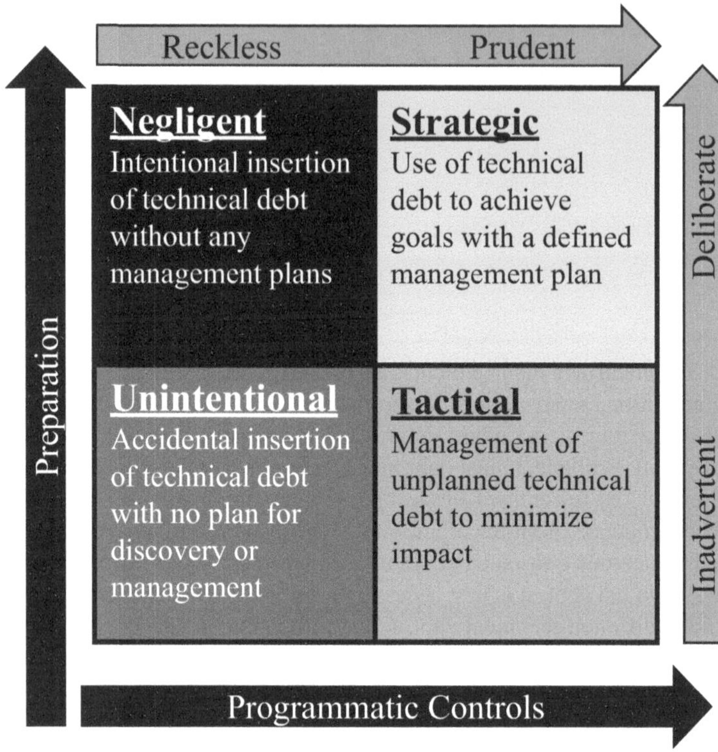

Fig. 7.1 Technical debt quadrant with controls

The first two parts of this book focused on the preparation side of the process. By understanding the origins and definitions of technical debt and its application to the system lifecycle, you are now prepared to deal with technical debt within your system. In this part of the book, we introduce the programmatic controls necessary to execute technical debt management within system development.

To understand the programmatic role of technical debt management, let's return to the technical debt context map, shown in Fig. 7.2. Technical debt management appears on the right side of the map and manages technical debt items to reduce the long-term consequences. Therefore, our technical debt management process should not be focused on preventing technical compromises. Instead, it needs to provide tools and techniques to limit the negative long-term impacts of those technical compromises.

In this chapter, we introduce the technical debt management cycle and provide comparisons to typical risk management processes. We also introduce the technical debt register, which is the primary tool used to track technical debt within a program. Finally, we provide some methods to help reduce the occurrence of unintentional technical debt.

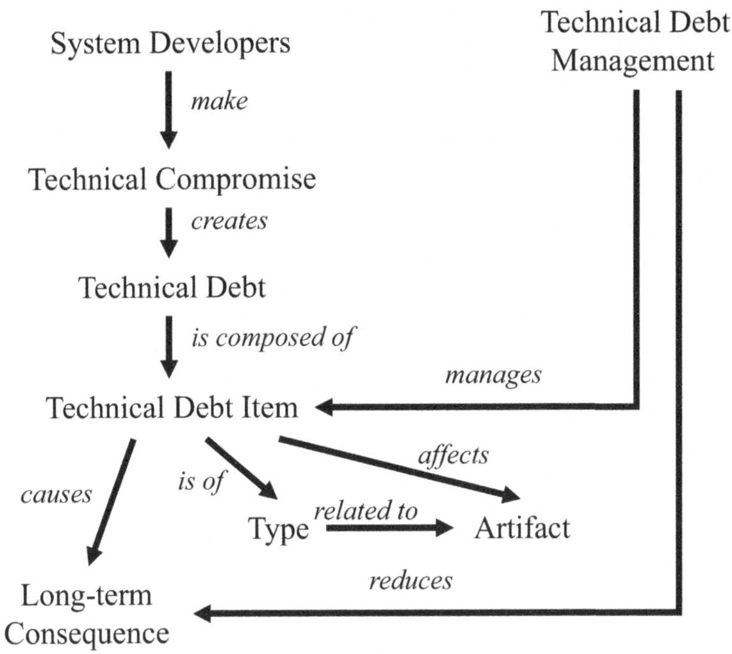

Fig. 7.2 Technical debt management reduces long-term consequences

7.1 Technical Debt Management Cycle

Technical debt can be inserted into a system at any time, by any developer. No approvals are needed, and the developer may not even be aware that they are introducing the debt. This aspect is one of the challenging features of developing a technical debt management process—how do you manage something that you cannot control?

To answer this question, we need to develop a cycle that consists of both discrete and continuous activities. The discrete activities apply to a single technical debt item as it moves through its lifecycle. The continuous activities are always applied to the system to identify new technical debt items. Fig. 7.3 graphically depicts the technical debt management cycle, which consists of five processes: identification, measurement, prioritization, repayment, and monitoring [1]. We provide a brief overview of each process here and then discuss each one in more detail in the subsequent chapters.

The identification process seeks to find technical debt within the system. There are three types of technical debt identification: proactive, active, and reactive. Proactive identification assesses the potential consequences of a technical compromise prior to it being made. Active identification identifies technical debt as it is inserted into the system, especially during implementation. Reactive identification looks at the

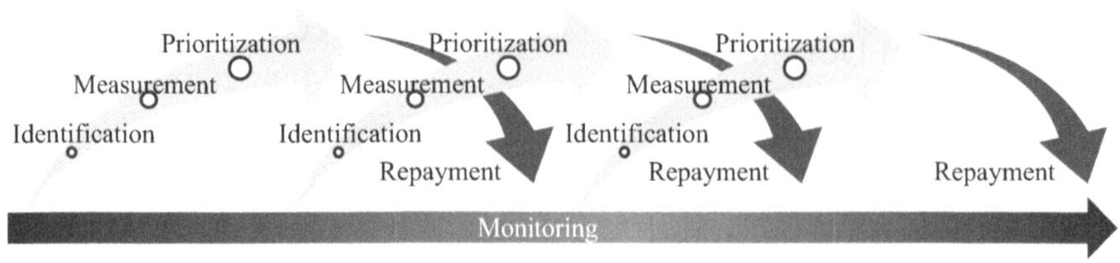

Fig. 7.3 Technical debt management cycle

system and attempts to determine technical debt that has already been incurred, perhaps unintentionally.

The measurement process quantifies the amount of technical debt. This step enables prioritization and establishes the debt exposure of the system. Quantifying the principal, taxes, interest, and fees is a complicated task and will be system-dependent.

In prioritization, the system developer decides which technical debt items to repay and when to repay them. Prioritization is key to paying down debt by identifying the most impactful technical debt items and identifying when they can be repaid without disturbing the rest of the system development.

Repayment is the process of performing work to remove a technical debt item. This work can take the form of removing the original technical compromise or eliminating the long-term impacts of the compromise. For example, a system redesign may not be required to eliminate user fees if the performance can be increased through optimization of another part of the system. During repayment, the principal and taxes are repaid; interest and fees are paid during system development and use.

These four processes establish a sequence of events—once technical debt is identified, it must be measured, prioritized, and eventually repaid. This sequence is repeated as new technical debt items are found. Outside of this sequence is the monitoring process. Monitoring focuses on the continuous observation of technical debt in the system—both the debt that is already established and potential new sources. The monitoring process tracks the current debt load to determine if interest is becoming more likely to come due or if technical fees are increasing in frequency. This process is critical to preventing the technical debt burden from becoming excessive on the system.

7.2 Comparison to Risk Management

In Sect. 1.3, we briefly touched on the difference between technical debt and risks. Specifically, we highlighted that:

- Risks are ***uncertain events*** that, if they occur, ***have a negative effect*** on the system [2]
- Technical debt occurs because of an ***event which occurred*** (the technical concession) but the ***magnitude of the effect is uncertain***.

It may appear convenient to combine technical debt management with existing risk management processes. After all, there is a probability and consequence associated with technical debt, just like in risk management, and technical debt items need mitigation steps to reduce their impact. While this approach is better than not managing technical debt, technical debt has enough unique characteristics that make it better suited to a separate, but connected, management process. In this section, we review risk management processes and establish the relationship between risk management and technical debt management.

7.2.1 Overview of the Risk Management Process

The risk management process is a continuous process by which the system developer identifies, assesses, treats, and monitors risks in the system [2]. Risks are often characterized by their likelihood and consequence and displayed on a 5 × 5 matrix, as shown in Fig. 7.4. The likelihood is the probability that the risk triggering event occurs, and the consequences are the negative impacts on the system, which can occur in any of the system dimensions, resulting in increased costs, schedule delays, or reduced performance. Proper risk management ensures that the system developer has a plan to minimize either the likelihood of a risk occurring or the consequences if it does occur, or both. Once the risk triggering event occurs, the risk becomes an issue, and the system developer must deal with the consequences.

Risks are often monetized to allow comparison. If technical and schedule consequences can be converted to dollar amounts, then risks with consequences in different areas can be compared. The risk exposure is calculated by determining the financial impact of the consequence and then multiplying it by the likelihood. For example, a risk with a $1 million consequence and a 10% likelihood would be valued at $100,000. Project management can use these risk numbers to calculate the overall

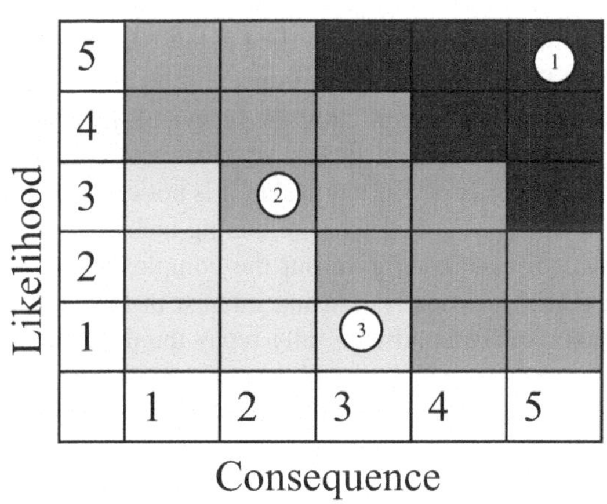

Fig. 7.4 Notional risk matrix

risk exposure and hold sufficient project budget in reserve to cover the risk. The risk management process is executed at the Risk Management Board (see Sect. 8.2.1.2).

Risk mitigation strategies focus on reducing the likelihood of the risk trigger event or on reducing the consequence of the risk if it occurs (or both). The developer may take steps to minimize the risk of a part failure occurring by improving defect screening prior to part installation. At the same time, the developer may reduce the consequence of the part failure by including redundancy in the system design.

7.2.2 Technical Debt Management Versus Risk Management

Unlike risks, whose likelihood is uncertain, technical debt is introduced into the system through the occurrence of a known event. Technical debt is the result of an action on the part of the system developer (either intentionally or unintentionally) to make a technical compromise, whereas risks are identified prior to the occurrence of the trigger events. The occurrence of technical debt, therefore, is within the control of the system developer and the system stakeholders, which allows technical debt management to be proactive—addressing and evaluating the technical compromises before they are made.

The other difference between risks and technical debt is the time span between the event occurrence and the consequences. The trigger event of the risk typically causes immediate consequences. With technical debt, the technical compromise may create impacts which are not observed for a long time or which may never be observed. Therefore, the probability of a technical debt item is not associated with whether or not the debt will be incurred (the principal), but rather with the potential for future impacts on the health of the system (the interest). Technical debt has a probabilistic consequence instead of a probabilistic likelihood.

Since the technical compromise is certain, technical debt mitigation strategies cannot focus solely on reducing the likelihood of the technical debt entering the system. Part of technical debt management is training staff to recognize and limit the insertion of technical debt (see Chap. 8). However, most mitigation strategies will focus on limiting the impact of the long-term consequences, either by repaying the principal (eliminating the technical debt) or by mitigating the development of interest and fees. To use a software example, consider a software module that is "spaghetti-code": it works, but it is not easy to understand or debug. This module is a primary source for compounding technical debt, as every module that interfaces with it needs to figure out the complexity and the sequence of function calls to execute it properly, causing interest in the form of longer development times and other workarounds. To fully repay the debt, the system developer needs to rewrite the module, which would be time-consuming and at risk of errors due to the code complexity. One mitigation step for this technical debt item is to contain the module behind a wrapper module. All interactions would then run through the wrapper, which can be cleanly constructed with well-defined functions. While the wrapper would slightly decrease performance due to the extra function calls, it contains the

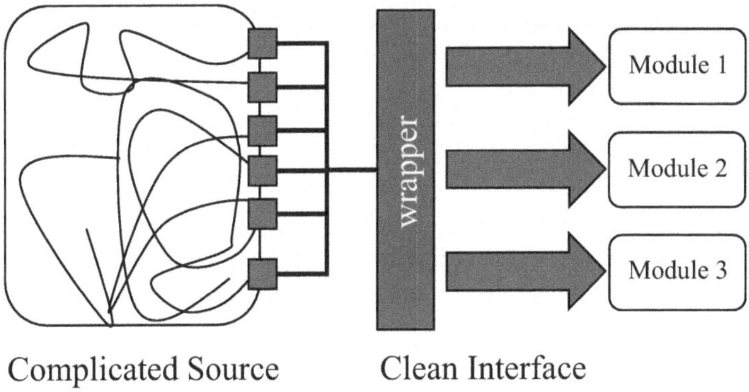

Fig. 7.5 Implementation of wrapper class

technical debt to the single module and prevents its spread. Then, when time allows, the debt-ridden module can be combined with the wrapper module to fully remove the technical debt. This architecture is shown in Fig. 7.5, where complicated source code is wrapped to create clean interfaces to each of the three modules. Implementing the wrapper is paying a smaller interest amount to mitigate the further spread of the technical debt. However, this implementation creates taxes that would need to be paid if and when the wrapper is removed.

Similar to risk management, technical debt should also be managed through a board process, where new technical debt items are proposed, categorized, and prioritized. The technical debt review board, discussed in Chap. 11, tracks the progress of technical debt mitigation and forces the system developer to execute the technical debt management process.

7.2.3 How Risks Become Technical Debt

While technical debt and risks are different, they are related. Even though risks can be mitigated, the likelihood of the triggering event, or its consequence, may not fully disappear. At this point, the system developer may choose to accept the risk—further actions to reduce the likelihood or consequence are no longer worth the effort. When a risk is accepted, it must be assessed for the introduction of technical debt.

Accepted risks that create technical debt are those risks that introduce a technical compromise. Let's consider an example of how this occurs. A system developer decides to contract out the development of one of the components of their final system. The system developer carries a risk on the quality of the parts delivered by the contractor and decides to mitigate the risk by testing engineering units produced by the contractor. The tests reveal that a small percentage of the units underperform. However, due to schedule and budget constraints, the system developer decides that the risk is acceptable. Accepting the risk has introduced both fee-bearing and interest-bearing technical debt. If the part fails during operations, then the delivered system will underperform. If the part fails during integration and testing, then the developer will have longer development timelines due to the need to replace the part and retest. The technical compromise has been made: the risk of failure is accepted. The consequence is uncertain: the part will fail with some probability, and the impact depends on when the part fails and how many other areas interface with that part.

Accepting any risk to the performance of the system may create technical debt, regardless of whether the consequence is in the cost, schedule, or performance dimension. Negative impacts in any of these dimensions impact the overall health of the system, and if the risk event occurs, it will create a negative impact. Accepting a risk to the cost or schedule dimension does not create technical debt as there is no associated technical compromise. We'll discuss how accepting risks creates technical debt in more detail in Sect. 8.2.1.2.

7.2.4 How Technical Debt Becomes Risks

The relationship between technical debt and risks goes both ways—by accepting technical debt into the system, the system developer can also introduce risks. Let's return to the operating system example from Sect. 3.4. In this example, the system developer took on technical debt when they selected to use an old operating system to save the development schedule. This operating system is more susceptible to cyber-attacks than the newer version. The work required to upgrade the operating system, including principal and taxes, would be tracked as part of the technical debt item. The interest in this technical debt item is related to the additional work required to work around the old operating system: updating new applications, performing additional cyber patches and hardening, and reduced user performance.

What the technical debt item does not track is the increased likelihood of a cyber-attack on the system. The technical debt item accounts for the increased work to mitigate the vulnerabilities (the increased rate of patches and hardening), but it does not account for the consequences of an attack. Instead, the technical debt item has introduced a new risk into the system—the increased likelihood of a successful cyber-attack. The risk statement could be written as:

> Given that the old version of the operating system is used, if a hacker attacks the system using known exploits, then they will gain access to critical functions.

The use of the older operating system has increased the likelihood of a cyber-attack being successful. However, a cyber-attack occurring is still an uncertain event, which indicates that the consequences of the attack should be treated as a risk and not as technical debt.

7.3 Technical Debt Register

Before delving deeper into the technical debt management process, we first need to discuss the technical debt register. A technical debt register is a tool, like a risk register, that can be used to track and manage technical debt within a project [3]. Each technical debt item identified within the project is tracked in the register until it has been repaid. The register contains enough information about each technical debt item to enable the system developer to make prioritization decisions regarding the schedule for repaying the technical debt and whether a particular technical debt item should even be repaid. When viewed in total, the technical debt register provides insight into the overall technical debt load of the project. Various techniques can be used to create the register, such as Jira[1] or a simple spreadsheet [4]. Organizations should use a technical debt register format and tools that easily integrate into their current workflows.

The technical debt register should be reviewed regularly, often in conjunction with the risk register, and developed and maintained throughout the project lifecycle as the source of accumulated technical debt. Its regular review is critical to determine when to repay technical debt items, changes in the interest probability or amount (either increasing or decreasing), and the current project debt exposure.

In the next sections, we will review content that should be included in the technical debt register and discuss how the technical debt register can be analyzed to give insights into issues with system development. These sections represent proposed content, but each system developer should tailor the technical debt register to meet their needs.

7.3.1 Technical Debt Statement

The first item to consider in the technical debt register is the technical debt statement. The technical debt statement needs to concisely convey the critical aspects of the technical debt item, such as the short-term benefit, the compromise, and the potential long-term impacts. It should provide enough information for a reader to understand the problem but should not go into details and descriptions of the compromise or the consequences. The goal of the statement is to enable concise discussion of the technical debt item and its resolution. Additional details will be contained in the other fields within the technical debt register.

We can leverage concepts from well-formed risk statements to create a syntax for the technical debt statement. A well-formed risk statement includes the condition and consequences of the risk. If the cause of the risk is known, then it should be included as well [6]. Including these items leads to risk statement formats like:

[1] Jira, produced by Atlassian, is a workflow management product that allows users to enter tasks, view and manage task status, and generate reports. While initially design for use in Agile software development, it has expanded into multiple fields and domains [5].

If (condition), then (consequence)

Because (cause), (condition), resulting in (consequence)

In each of these formats, the parenthetical terms are replaced with the actual values. A risk of faulty parts from a supplier would be captured as:

If the parts received from the supplier require rework, then the start of testing will be delayed by two weeks.

Because the supplier has poor quality control, the parts received from supplier a may require rework, resulting in the start of testing being delayed by two weeks.

We define a similar format for the technical debt statement, leveraging the critical components of a technical debt item: the short-term benefit, the technical compromise, and the long-term consequences. If the technical debt is unintentional, then the benefit may be unknown. Therefore, there are two potential technical debt statement formats:

To (short-term benefit), (compromise was made), resulting in the potential for (long-term consequence).

Unintentionally, (compromise) was made, resulting in the potential for (long-term consequence).

The statement starts with the short-term benefit, which is the reason why the technical debt was incurred. For intentional technical debt, this benefit should be clear: to maintain schedule, to prevent cost growth, or to achieve performance in another area. For unintentional technical debt, which must be discovered after its insertion, the short-term benefit may not be as clear. In cases where the short-term benefit is unknown, the opening clause is replaced with "unintentionally" or can be removed entirely. In the technical debt statement, the compromise should be short but still well defined, indicating what aspect of system performance was traded to achieve the benefit. Finally, the potential long-term consequences are identified. The consequences should indicate the affected artifacts and the associated impacts.

Let's return to the previous risk. The supplier is known to have poor quality control. As part of the mitigation plan, the system developer works with the supplier to increase the quality control processes and to better screen out defective parts. These actions decreased the likelihood of the supplier delivering a defective part from 50 to 10%. The system developer decides to accept this risk as further mitigation efforts would be too costly. This risk acceptance introduces technical debt into the system (see Sect. 7.2.3). The corresponding technical debt statement would be:

To maintain the project budget, a 10% likelihood of defective parts from the supplier was accepted, resulting in the potential for the start of testing to be delayed by 2 weeks.

You may be reading this statement and saying that it is the same as the risk statement—there is a potential delay of two weeks in the system. This is true! However, the difference is that no further steps are being taken to mitigate the risk of the supplier delivering poor-quality parts. The system developer has made the technical compromise of accepting a higher-than-desired defect rate. The consequences are delays to the overall system schedule, but these consequences may or may not happen. Since the risk has been accepted, the impacts should be treated as technical debt—there is no longer a triggering condition for the risk.

7.3.2 *Formatting the Technical Debt Register*

While the contents of a technical debt register can be tailored by each project, we suggest dividing the entries for each technical debt item into three sections: descriptive fields, management fields, and measurement fields. The descriptive fields describe the compromise that was made and its impact on the system. The management fields define the state of repayment and prioritization. The measurement fields quantify the technical debt item. Table 7.1 shows a set of recommended fields for a technical debt register, and an example register can be found in Appendix B.

7.3.3 *Using the Technical Debt Register*

Once the technical debt register is created, it becomes the primary tool used to manage the system technical debt. The register provides the authoritative list of technical debt items present in the system and the information needed to assess the balance and total cost of these items.

The technical debt register is a living document and needs to be updated at technical debt review boards (see Sect. 11.1) and anytime new technical debt items are identified. The items on the register should be regularly reviewed to determine if the mitigation steps are successful and if there are any changes to the interest amount or probabilities. With new releases, the technical debt register should be adjusted to reflect any items that have been repaid.

The register provides a mechanism to sort technical debt items by priority, principal, or other fields. This sorting allows the system developer to determine which work to include in the next development cycle. Sorting by artifact enables an assessment of which components are most in need of repair or are most likely to cause problems, allowing the developer to prioritize resources in repaying technical debt.

We will discuss using the technical debt register throughout the technical debt management processes defined in the following chapters.

7.4 Reducing Unintentional Technical Debt

Intentional technical debt is traceable directly to the pressure from stakeholders, and reducing this pressure is largely a project management activity related to stakeholder management and is outside the scope of this book. However, unintentional technical

Table 7.1 Recommended technical debt register fields

Category	Field	Definition
Descriptive	Identifier	The unique identifier for the technical debt item
	Technical debt statement	The statement describing the technical debt item, as defined in Sect. 7.3.1
	Description	A free-text field to provide more information about the technical debt item. This helps to add context around either the rationale for the making the compromise, the impacted artifacts, or potential long-term consequences. The description should include the cause (why the technical debt item was inserted) and the long-term consequences
	Type	The type of technical debt (see Chap. 4)
	artifact(s)	The artifact or artifacts associated with the technical debt item. The artifact list should include those affected by the principal, taxes, interest, and fees
Management	Status	The current status of the technical debt item, using project-specific fields such as "Accepted," "Mitigating," etc.
	Prioritization	A project-dependent prioritization for repaying or mitigating the impacts of the technical debt item. This field helps with resource allocation to technical debt repayment
	Mitigation steps	The actions that need to be taken to reduce the long-term impacts of the technical debt
Measurement	Principal	The measured amount of the principal (see Sect. 9.1)
	Taxes	The estimated amount of the rework required to correct components when the principal is repaid (see Sect. 9.3)
	Interest amount	The measured amount of the interest amount (see Sect. 9.2.2). For technical debt items that incur fees, this field is not used
	Interest probability	The likelihood that the technical debt item will need to be repaid within a specified period of time (see Sect. 9.2.1). This field can help with the prioritization of the technical debt items. For technical debt items that incur fees, this field is not used
	Fee amount	The amount of fee paid by the user (see Sect. 9.4). For interest bearing debt, this field is not used
	Fee frequency	How often a user will have to pay the technical fee. For interest bearing debt, this field is not used

debt is due to the development team, and therefore we provide some strategies for reducing its occurrence.

In order for system developers to avoid unintentional technical debt, they must know what technical debt is and must be aware of how to manage it. Therefore, it is necessary to educate system developers about technical debt and introduce the technical debt lexicon into the system development processes. Once the team is thinking about technical debt, active measures can be taken to reduce the occurrence of unintentional technical debt.

One method to reduce unintentional technical debt is to increase communication across the team. When a developer works in isolation, or a component team is isolated, then they will make decisions that enhance their specific areas without consideration for overall system impact. These decisions are not made with the intent to harm the rest of the system, but may do so because of a lack of knowledge. Therefore, it is essential that design decisions are communicated and reviewed across the greater team.

One approach to increasing communication is concurrent engineering. Concurrent engineering is the process by which engineering processes are executed in parallel with frequent communication between engineers. Increased communication reduces the potential for technical debt driven by technical trades by enabling real-time analysis of the consequences of a trade in the context of the mission. Properly sequencing tasks by tracing dependencies, through tools such as design structure matrices, reduces the risks of rework due to out-of-order development [7].

Other approaches to minimizing unintentional technical debt include more frequent reviews—either via peer reviews or via sprint reviews in projects using agile and adaptive development strategies. These reviews, discussed in more detail in the next chapter, enable early detection of unintentional technical compromises before the interest on the debt grows too large.

Another source of unintentional technical debt is the over-parallelization of tasks. System developers will often parallelize activities to decrease the overall development timeline. However, doing so introduces technical debt into the system—if one side of the interface changes, all components that use that interface will also need to change. Therefore, to reduce this technical debt, it is essential that critical interfaces are documented and baselined before starting dependent tasks.

7.5 Summary

In this chapter, we introduced the technical debt management cycle, which we will elaborate on in the following chapters. Technical debt management shares a lot of similarities with risk management, and the processes should be connected, as accepting risks can generate technical debt. We introduced the technical debt register, a key tool in managing technical debt. While we provided suggested formats and contents for a technical debt register, each system developer should tailor the technical debt register to best meet their needs.

We concluded the chapter with a discussion on how to reduce unintentional technical debt. This type of technical debt is especially dangerous to the system, as the developer may not know it exists. The technical debt management cycle aims to identify technical compromises before they introduce technical debt, as well as to identify existing technical debt in the system, which may have been inserted unintentionally. The identification process is the subject of the next chapter.

References

1. Li, Z., Liang, P., & Avgeriou, P. (2014). Architectural debt management in value-oriented architecting. *Economics-Driven Software Architecture*, 183–204.
2. Walden, D. D., Roedler, G. J., Forsberg, K. J., Hamelin, R. D., & Shortell, T. M. (Eds.). (2015). *Systems Engineering Handbook* (4th ed.). John Wiley and Sons.
3. Rosser, L. A., & Norton, J. H. (2021). A Systems Perspective on Technical Debt. *IEEE Aerospace Conference (50100)*, 1–10. https://doi.org/10.1109/AERO50100.2021.9438359
4. Kruchten, P., Nord, R., & Ozkaya, I. (2019). *Managing Technical Debt: Reducing Friction in Software Development*. Addison-Wesley Professional.
5. Atlassian. (2026). *Jira*. Atlassian. Retrieved February 22, 2026, from https://www.atlassian.com/software/jira
6. Thompson, J., & Stump, S. (2017, November–December). How to Write a Good Risk Statement. *Defense AT&L Magazine*. https://www.waru.edu/sites/default/files/Migrate/DATLFiles/Nov-Dec_2017/Thompson_Stump.pdf?Web=1
7. Avnet, M. S., & Weigel, A. L. (2010). An application of the Design Structure Matrix to Integrated Concurrent Engineering. *Acta Astronautica*, 66(5–6), 937–949.

Chapter 8
Identifying Technical Debt

Abstract The first process in the technical debt management cycle is identification: the means by which technical debt items are identified in the system. Methods are introduced to identify technical debt proactively (before the technical compromise is made), actively (at the time of the technical compromise), and reactively (after the technical compromise has already been made). These methods focus on tailoring existing systems engineering reviews and processes to include technical debt identification techniques. The technical debt timeline is also provided, demonstrating how technical debt can initially benefit system development prior to causing long-term consequences.

Keywords Technical debt identification · Technical debt timeline · Process and review board tailoring · Digital engineering

The first process in the technical debt management cycle is identification. As depicted in Fig. 8.1, the goal of the identification process is to find technical debt items–either at the point of insertion, for intentional technical debt, or through a review of the state of the system, for unintentional technical debt. Identifying technical debt requires modifying existing systems engineering processes to include a specific focus on technical debt identification. In this chapter, we provide methods to proactively, actively, and reactively identify technical debt such that it can be effectively managed.

8.1 Technical Debt Timeline

To understand how we can identify technical debt within a system, we must first understand the timeline of technical debt items once they are in the system, which is shown in Fig. 8.2. In this figure, it is important to note that the time axis includes milestone-based time points and does not represent actual duration. Each technical debt item in each system will have a unique time scale.

H. Kleinwaks, *Managing Technical Debt in Systems Engineering*,
https://doi.org/10.1007/978-3-032-30360-8_8

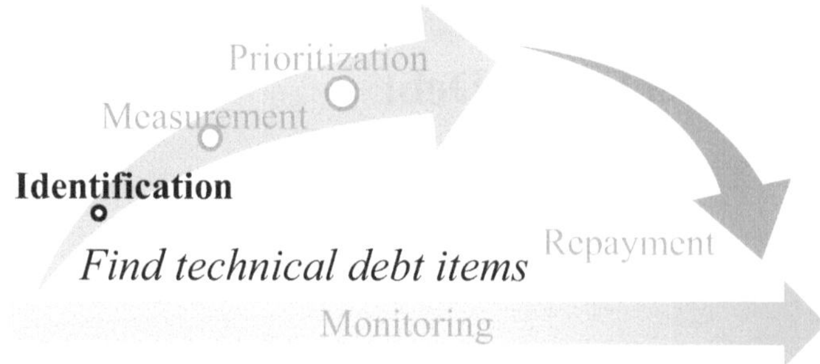

Fig. 8.1 Identification process

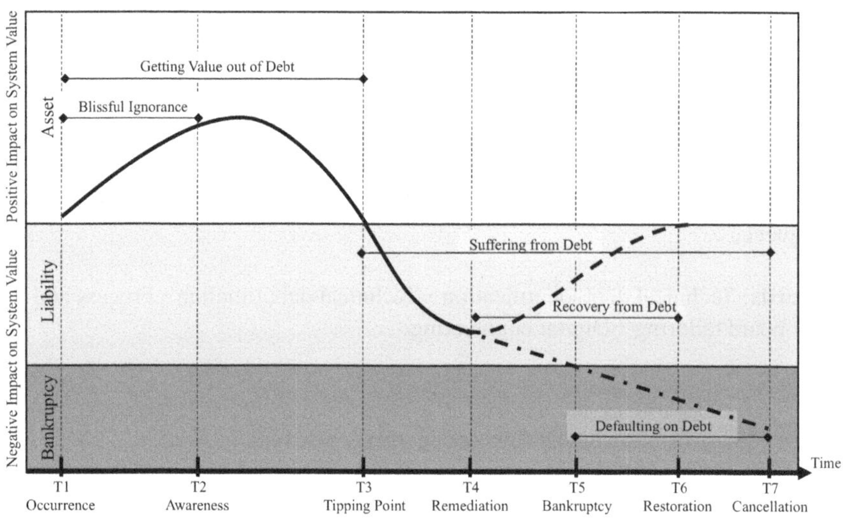

Fig. 8.2 Technical debt timeline [1, 2]

As we've noted previously, there are times in the system lifecycle where technical debt can be beneficial to the system. After all, technical debt gets introduced to provide a short-term benefit. The top section in Fig. 8.2 represents the time when the technical debt is an asset. The technical compromise has a positive impact on the system value: it accelerated a release, lowered cost, or enabled additional development to begin.

If the technical compromise was unintentional, then the system developer starts in a state of 'blissful ignorance'–the time when the technical debt item is beneficial to the system and the developer is unaware of its presence. This time, starting at T1—Occurrence, lasts until T2—Awareness when the technical debt is identified. For intentional technical debt, T1 and T2 are the same time—the system developer is aware of the technical debt item when it occurs. Being aware of technical debt does not mean that it stops being of value to the system developer and the system developer continues to get value out of the technical debt item. However, as interest and taxes accumulate, the net positive value decreases until T3—Tipping Point is reached [2].

At the tipping point, the negative impacts on the health of the system start to outweigh the short-term benefits, pushing the system into the middle section of Fig. 8.2. The technical debt item has become a liability to the system developer. At this point, the system is suffering from the technical debt, and will continue to do so until the debt is either repaid and the system value is restored, or the project is cancelled.

At T4—Remediation, the system developer, aware of the technical debt and its negative impacts, has a decision to make. They can either start to repay the technical debt and recover, moving to T6—Restoration, or they can let the technical debt remain untouched, which can eventually lead to T5—Bankruptcy and T7—Cancellation, in the bottom section of Fig. 8.2. We will discuss repaying technical debt in Chap. 10 and technical bankruptcy in Chap. 12. Note that the developer can choose to start repaying the technical debt at any point in the timeline, including prior to the tipping point and after bankruptcy, but these paths are not shown. The stages and gates of the technical debt timeline are summarized in Table 8.1.

The technical debt timeline tells us that technical debt items may eventually create liabilities in the system, but that they can also be assets. If properly identified, the duration between gates T2—Awareness and T3—Tipping Point can be maximized. However, the ability to do so is dependent upon the technical debt item and how it is inserted in the system. To understand this point, we can take the four technical debt quadrants from Fig. 2.2 and map them onto a notional technical debt timeline to produce Fig. 8.3. Note that Fig. 8.3 only shows technical debt which is repaid and not that which goes into default.

Negligent debt is reckless—it is inserted into the system without a plan to repay it. It is also intentional, which means that there is no "blissful ignorance." The technical debt is known when it is inserted, and therefore the timeline starts by getting value out of debt. Unintentional debt does start with the "blissful ignorance" phase, and this phase can be quite long as it may be a significant amount of time before the technical debt is identified. However, the difference between the awareness point (T2) and the tipping point (T3) may be quite short, as the technical debt is typically only found once it has started causing problems. The suffering stage for reckless technical debt is longer than that associated with prudent debt. The system developer must come up with the plan to mitigate the technical debt while suffering from it and the likelihood of reaching bankruptcy is substantially higher.

Strategic technical debt is knowingly introduced into the system with a plan to repay it and starts at the getting value stage. There is no time of "blissful ignorance," since the technical debt is identified at the point of insertion. However, the getting value stage is longer since the introduction of the technical debt is accompanied by a plan to mitigate its impacts and repay it over time. Similarly, the suffering and recovery time periods are shorter since they are planned out in advance. Tactical technical debt starts with a short "blissful ignorance" phase, since tactical debt is often unintentional—it is a prudent reaction to unexpected conditions. This phase is

Table 8.1 Stages and gates of the technical debt timeline, adapted from [1]

Stage	Definition	Entry gate
Blissful Ignorance	Technical debt has unintentionally been inserted into the system and the system developer and/or the stakeholders are receiving the short-term benefits. Long-term impacts on the system health have yet to appear. This stage only applies to unintentional technical debt	T1—Occurrence: the unintentional making of a technical compromise
Getting value out of debt	The technical debt is an asset to the system developer and/or stakeholders. The system developer may or may not be aware of the technical debt, yet they are receiving the short-term benefits without experiencing the long-term impacts on the system health. For unintentional technical debt, this stage overlaps with the Blissful Ignorance stage. For intentional technical debt, this stage starts with the making of the technical compromise As the long-term impacts of the technical compromise accrue, the value delivered to the system of that compromise decreases, eventually moving the technical debt from an asset to a liability	T1—Occurrence/ T2—Awareness: the insertion and identification of the technical debt. For intentional technical debt, T1 and T2 overlap
Suffering from debt	The technical debt has become a liability and the long-term impacts on system health outweigh the short-term benefits. The system developer pays interest and the system user pays fees. This stage lasts until the technical debt is fully repaid or the project is cancelled	T3—Tipping Point: the point at which the technical debt switches from an asset to a liability
Recovery from debt	The system developer initiates the repayment of the technical debt by paying the principal and taxes. This stage includes items such as retesting hardware systems, completing interface definitions, and refactoring code. This stage lasts until point T6—Restoration, where the performance dimension is restored as much as possible. The system still suffers from the technical debt until recovery is complete	T4—Remediation: the start of the repayment plan
Defaulting on debt	If the system developer does not repay the technical debt item, then they may eventually reach bankruptcy, where development cannot continue until the technical debt is repaid. In this stage the developer has defaulted on their debt and cannot complete other work until the debt item is resolved. If the developer cannot repay the debt, then the project will eventually reach T7—Cancellation	T5—Bankruptcy: the state where the developer must repay the technical debt to continue in the system lifecycle

short because the developer rapidly develops a plan to manage the technical debt. In both strategic and tactical debt, the getting value phase is longer than the suffering phase because of the additional planning associated with prudent technical debt.

Figure 8.3 makes it clear that the duration between the tipping point and the awareness point is maximized for prudent technical debt and especially for strategic technical debt. However, technical debt cannot always be identified at the point of

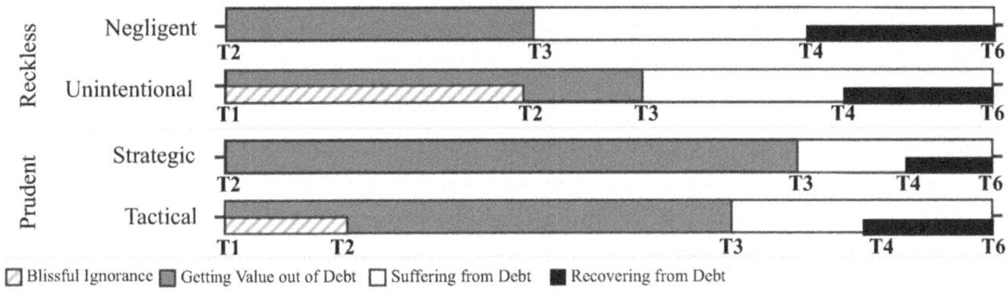

Fig. 8.3 Notional timeline of technical debt occurrence by group [1]

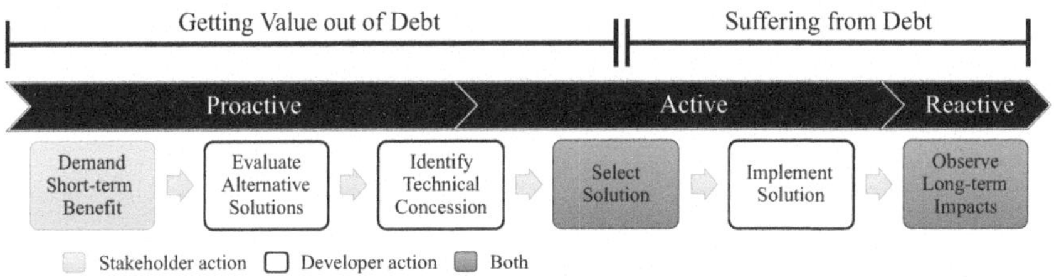

Fig. 8.4 Approaches to technical debt identification in sequence [1]

insertion, and sometimes must be identified once it is already in the system. Figure 8.4 shows a time-based view of the sequence by which technical debt is created. The later the technical debt item is identified, the larger the cost to repay the technical debt [3] and the more likely that the system was suffering from the technical debt item instead of getting value out from it. The arrows identify the three different approaches to technical debt identification, and where they apply in the sequence of technical debt generation.

Initially, system stakeholders will induce pressure on system developers such that they can achieve a short-term benefit, such as an earlier release or a cost savings. The system developer evaluates alternate solutions for providing the benefit and identifies any associated technical concessions. At this stage of the process, the system developer can proactively identify technical debt—when evaluating technical compromises associated with alternate solutions they can predict any associated technical debt.

After selecting the solution, the system developer will implement the solution and enters the phase of active technical debt identification. As the developer is making changes to the design or implementation, they may identify technical debt that was missed during the proactive design phase, or be forced into new technical compromises to achieve the desired outcome. This technical debt is identified at the point of insertion and it is critical that it gets logged into the technical debt register as soon as it is discovered, such that it can be properly managed.

Finally, after the technical compromise is implemented, technical debt may be observed through decreases in the performance dimension. If it was not already identified, then we are in the reactive phase—forensic methods must be applied to determine the source of the technical debt and the affected artifact such that it can be mitigated.

Putting these concepts together, we see that we want to identify technical debt as early as possible to limit the cost of repayment and the negative long-term impacts. Doing so will increase the duration of the "Getting Value Out of Debt" phase, allowing us to extract more value from the technical debt by having a well-formed management plan. If this is not possible, then we must actively identify the technical debt as soon as possible to reduce the duration of the suffering phase.

In the following sections, we will discuss the three timelines for identifying technical debt within the system. Proactive identification methods discover the technical debt before the point of insertion into the system. Active identification methods identify the technical debt as the compromise is being made. Reactive identification methods discover existing technical debt that is already in the system, whether due to a review process or by the observation of long-term impacts.

8.2 Proactive Technical Debt Identification

Proactive technical debt identification involves understanding the consequences of compromises in the performance dimension at the time they are made. The system developer must focus not only on the short-term benefit of changes to the system development plan but also on the long-term impacts. By being proactive, plans for repaying the technical debt (see Chap. 10) can be made when the technical debt is incurred, resulting in strategic technical debt—that which is accepted intentionally by the system developer with a plan to repay it. This debt is the least dangerous to the continued success of the system and can generate the value prior to being repaid, as shown in Fig. 8.3.

8.2.1 Tailoring Systems Engineering Processes to Proactively Identify Technical Debt

Technical debt arises through changes to the system, as the result of a technical compromise. While it is not always possible to identify technical debt prior to its insertion, especially with unintentional technical debt, we can augment common systems engineering processes to be aware of technical debt creation. In the following sections, we will discuss how to tailor change control boards and risk management boards, although technical debt identification should be included in all project review boards. These processes address changes prior to their insertion into the system and therefore are primary methods to proactively identify technical debt.

8.2.1.1 Change Control and Engineering Review Boards

Changes are inevitable during the development of a system. However, these changes must be managed to ensure that the system delivers the desired product without incurring too much technical debt. As part of the configuration management process, system developers typically implement a change control board (CCB) [4].

The change control board is the central point to manage changes that are proposed throughout system development. Proposed changes are brought to the CCB to review whenever they represent a change to the technical baseline and need to be approved by the CCB prior to being implemented within the system. The CCB will evaluate the proposed changes for technical merits, requirements compliance, and impacts to the project performance and development [4]. In some projects, the technical changes are handled and approved by an engineering review board (ERB) [5] and cost and schedule changes are handled by the CCB, meaning a proposed change must clear both hurdles: the ERB for technical approval and the CCB for cost and schedule approval.

These review boards provide opportunities to proactively identify technical debt. When reviewing the technical impacts of the proposed change, the review board can look at any technical compromises that are being made and determine the long-term impacts to identify the technical debt item(s) that would be created. The technical debt is measured (see Chap. 10) and this measurement can be used as one of the decision factors in approving or rejecting the proposed change to the system baseline.

The need for a technical debt evaluation during a review board is especially important when the proposed change reduces the performance of the system. The developer may be tempted to accept the change if they are focused on the short-term—by accepting the change, the system does not need to be reworked and development can continue. However, this focus will insert technical debt into the system.

It may be acceptable to introduce this technical debt, but to reach that decision, the technical debt must be assessed. What is the impact on the ability to validate the system? Will the users see a reduction in performance? How significant is the impact? By answering these questions, the system developer can consider both the short-term benefits and the long-term impacts of accepting the change, allowing them to make a fully informed decision. At ERBs and CCBs on my spiral development programs, we use the simple quad chart shown in Fig. 8.5. Each ERB and CCB presentation must describe the technical compromise, the affected artifact, the short-term benefit, and the long-term consequences prior to acceptance of the change request.

If the decision is made to accept the change, then the developer must add the new technical debt item to the technical debt register and must develop a mitigation plan for long-term impacts. Following this process allows the developer to track the technical debt at the point of insertion. Associating the technical debt review with existing formal review processes provides a simple mechanism to proactively identify technical debt.

Affected Artifact	Technical Compromise
Short-Term Benefit	Long-Term Consequence

Fig. 8.5 Quad chart for technical debt discussion at ERB

8.2.1.2 Risk Management Board

Throughout system development, the developer must manage risks to the system. A risk is an event that, if it occurs, can negatively impact system development [6]. We have already discussed how risks are both similar to and different from technical debt (Sect. 1.3) and compared the risk management process to the technical debt management process (Sect. 7.2). In this section, we discuss the risk management board and how it can be used to proactively identify technical debt associated with risk acceptance.

The risk and opportunity management board (ROMB) reviews risks and assesses the developers' progress in mitigating existing risks [7]. At the ROMB, risk handling plans are reviewed and the risk can be formally accepted, meaning the likelihood and consequence of the risk have reached a level where further mitigation is not desired. This occurrence could be because the likelihood and consequence have been sufficiently mitigated or because there are no more possible mitigations steps. As described in Sect. 7.2.3, once the risk is accepted, it has the potential to create technical debt. Therefore, the ROMB provides an opportunity to proactively identify technical debt when risks are accepted.

Prior to accepting a risk, the ROMB should define the technical debt that would arise from risk acceptance. Accepting the risk then involves both risk approval and approval of the created technical debt, which should be inserted into the technical debt register for management. This process is shown in Fig. 8.6.

This risk management process starts with the identification of a risk (step 1), which includes not only determining the likelihood and consequences but also the trigger event: the event at which the consequences are realized. Risk mitigation steps apply until the trigger event occurs (step 2) at which point the risk becomes an issue—its likelihood of occurring has reached 100%, because the trigger event occurred. While executing the risk mitigation steps (step 3), the risk should be reviewed to see if it has either been fully mitigated (step 4) or if the residual risk is acceptable (step 5). If the answer to either of these questions is "no," then risk mitigation steps continue. If the answer is yes, then the system developer must determine if any technical debt has been created (step 7) and if that technical debt is acceptable (step 8). If no technical

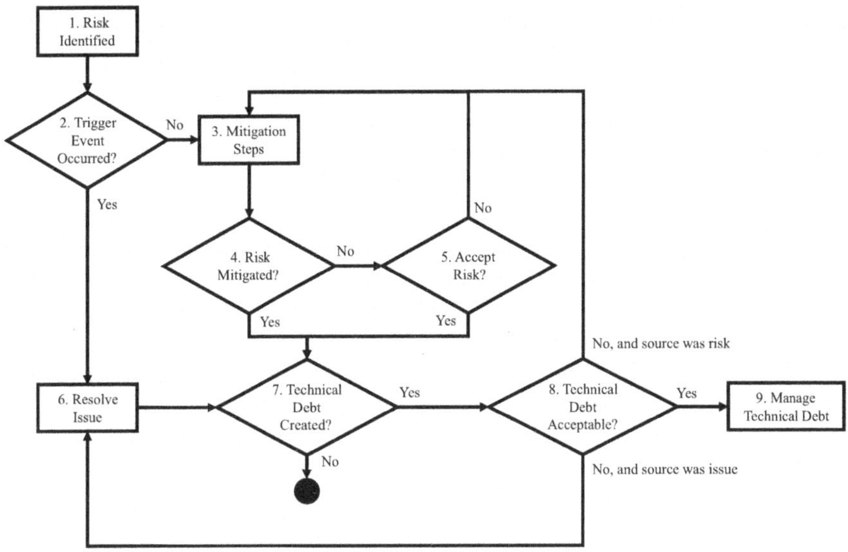

Fig. 8.6 Including technical debt in a risk management process

debt is created, then the process ends. If an unacceptable amount of technical debt is created then additional risk mitigation steps must be applied. If the amount of technical debt created is acceptable, then the resulting technical debt moves into technical debt management (step 9).

A similar process occurs once the risk trigger event occurs and the risk becomes an issue. The issue is resolved (step 6) and the resolution is reviewed for potential technical debt (steps 7–9), including a feedback loop to issue resolution if the resulting technical debt is unacceptable.

By including an analysis of technical debt creation along with risk and issue mitigation, the system developer can identify if the risk mitigation steps are sufficient or if they are making a technical compromise by prematurely accepting a risk. When the system developer accepts risks, they make a technical compromise. The assessment of technical debt concurrent with risk acceptance forces the early identification of technical debt and consideration of the long-term impacts of risk acceptance.

8.2.2 Predicting Technical Debt with Digital Engineering

We've discussed tailoring systems engineering processes to proactively identify technical debt. However, we've yet to explore how exactly technical debt can be predicted by reviewers. How do peer reviewers know that the document they are reviewing is making a technical compromise? An intentional technical compromise may be easy to identify, but detecting an unintentional compromise will be substantially harder. Predicting the impacts of design choices and implementations is complicated and often requires a system level view that is not shared by all members of the development team.

Digital engineering is an "integrated digital approach that uses authoritative sources of system data and models as a continuum across disciplines to support lifecycle activities from concept through disposal" [8]. Digital engineering enables the sharing of system-level knowledge across the development team. It moves communication between the development team from documents to digital models, enabling an integrated and consistent set of data artifacts that constitute the authoritative source of truth. The digital engineering framework includes the development of system models, analytical digital twins, and digital threads [9].

We use the term system model to refer to a digital model of the system structure and interactions, typically constructed through model-based systems engineering (MBSE) practices in a modeling language such as the Systems Modeling Language (SysML). These models connect the functional and structural components of the system and model use cases and system activities. Digital models move the source of information from traditional document formats (such as slide decks and text files) to an integrated system model. The integration of these data sources enables all system developers to access the same information from the same source and to always have access to the latest version of the data [10]. It must be noted that as these models grow, they become more complicated and can resemble complicated software programs, with package constructs, inheritance, and interdependencies, and relationships. Therefore, these models can be subject to the same technical debt problems as software programs [11].

Digital threads and digital twins are critical to being able to identify the impacts of technical debt throughout the system. Digital threads trace a path through the digital engineering framework, including requirements analysis, test and verification, and system performance, providing information to the stakeholder and system developers throughout the system life cycle. Once a technical compromise is identified that may affect an artifact, a digital thread traces that artifact and the associated compromise throughout the system lifecycle, identifying the affected components and the risk of compounding interest. Digital twins are realistic computer-based simulations of the system which use existing system performance information to predict future system behavior. Digital twins enable the analytical assessment of a technical compromise and its impacts on the system's figures of merit [9].

To demonstrate how a digital engineering framework can be used to predict technical debt, let's consider a hybrid sport utility vehicle. A system model for this vehicle is provided as a sample in the Cameo Systems Modeler 19.0[1] installation (hybrid sport utility vehicle.mdzip). This system model contains block definition diagrams, activity diagrams, state machines, and parametric models, allowing a full investigation of the system from requirements to design and performance. The model defines

[1] Cameo Systems Modeler, developed by Dassault Systèmes, is an MBSE environment that provides graphical interfaces to model systems using SysML. It allows users to display system diagrams, trace requirements, and execute state machines [12].

several requirements related to the performance of the hybrid SUV, including acceleration, fuel economy, and braking. This level of detail allows us to assess the impact of a technical compromise made on a lower-level part throughout the rest of the system.

Figure 8.7 shows the structural components that are connected to the wireless tire pressure monitor. If a technical compromise is made on the wireless tire pressure monitor, then the interconnected digital model enables rapid identification of connected components. These components are the ones that may be affected by the technical compromise. In this figure, we can see that any compromises related to the tire pressure monitor could affect the wheel assembly, chassis subsystem, power subsystem, and the entire Hybrid SUV. While some of these associations may have been intuitively obvious, some connections, like the power system, may not be. Therefore, the ability to automate the generation of potentially affected components provides a method to proactively identify technical debt throughout the system by allowing the developer to find all connected components and to evaluate the level of concern due to the compromise. Tools such as Cameo allow the generation of tabular data in addition to graphical data to enable additional processing and automated workflows.

Similarly, using a digital twin allows the system developer to estimate the impacts on the performance of the system from a technical compromise. The digital twin provides analytical models of the system that can be used to simulate the performance in different environmental conditions. If the technical compromise on the tire pressure monitor discussed above resulted in the acceptance of a tire pressure monitor with lower accuracy, then the effects of the lower accuracy on the overall performance of the car can be estimated. Tire pressure monitors provide indications to the driver when the pressure is 25% lower than the manufacturer's recommended pressure [13]. Figure 8.8 shows the effect of tire pressure on the fuel economy of a Toyota Corolla in 2014 [14].

While Fig. 8.8 does not show data for a hybrid SUV, similar trends would be expected. As the tire pressure goes down, the fuel economy decreases, with the largest effects seen at slow speeds. More significant drop offs are seen between 75% of the recommend pressure and 50% of the recommended pressure than between 75% of the recommended pressure and the recommended pressure. Therefore, a tire pressure monitor with reduced accuracy would not turn on at the specified threshold and would not notify the driver to refill the tire. The reduced pressure results in reduced fuel economy of the vehicle. If the sensor's accuracy was poor enough that it only turned on at 50% of the recommended pressure, then the driver could see a reduction in fuel economy of up to 10% at slow speeds. This reduction in fuel economy represents the long-term consequence of the compromise to accept the less accurate tire pressure monitor. The use of a digital twin to analyze the effects, as we did in this simple example, produces metrics that can be used to quantify potential impacts to stakeholders when deciding whether or not the make the technical compromise.

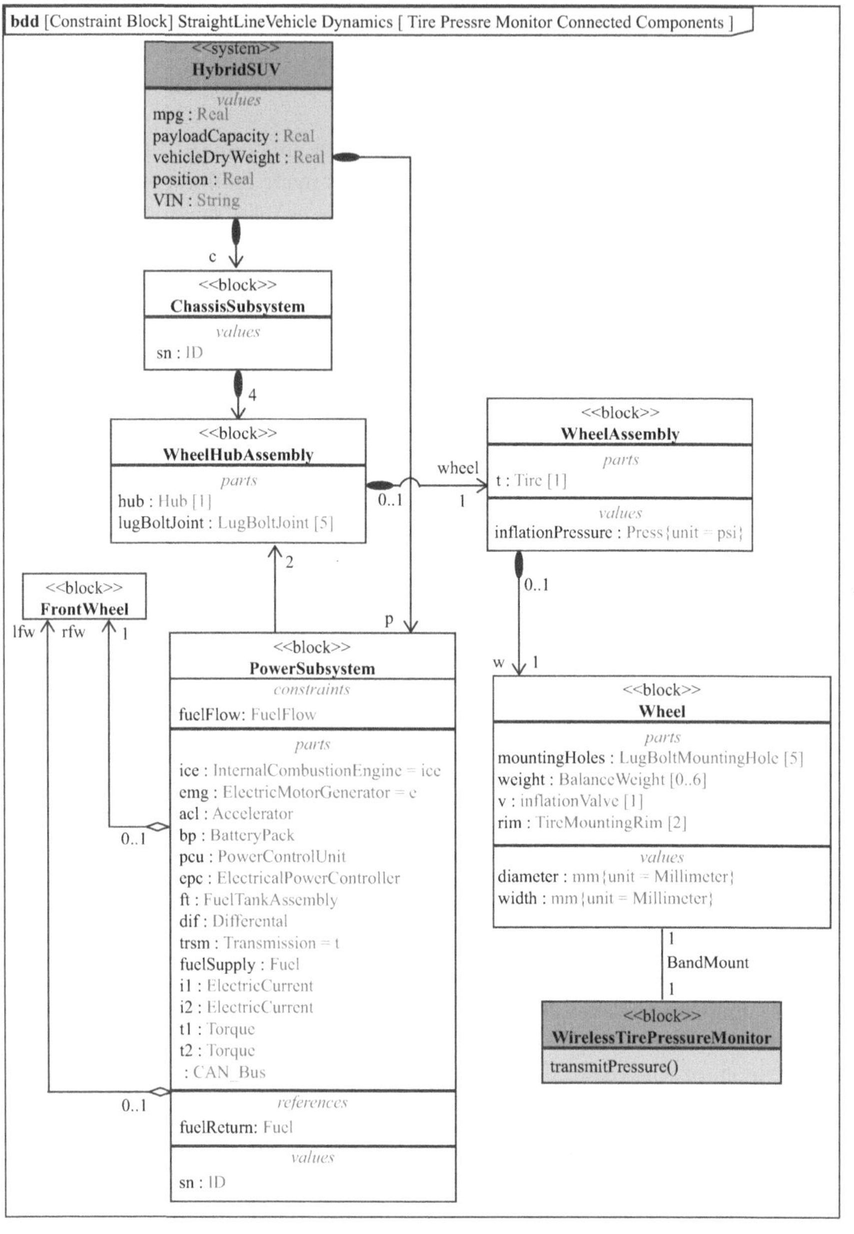

Fig. 8.7 Wireless tire pressure monitor connections

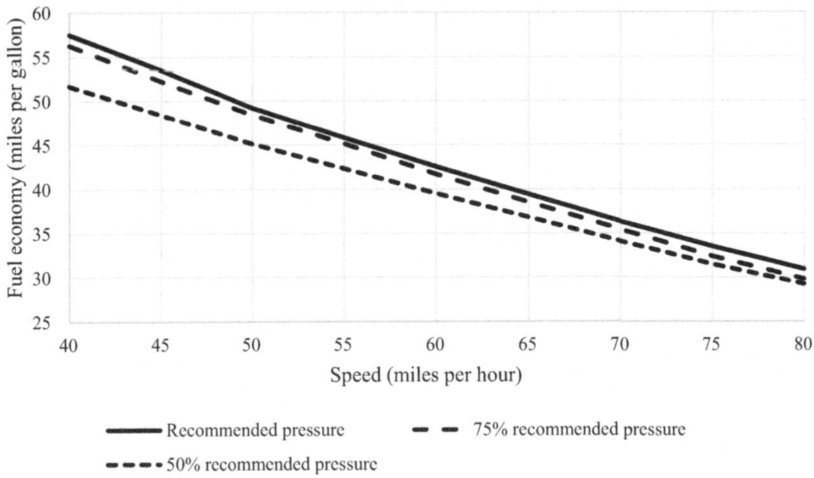

Fig. 8.8 Effect of tire pressure on fuel economy [14]

8.2.3 Predicting Technical Debt During Sprint Planning

Development teams using sprint-based strategies, such as Scrum, will plan out the work for a sprint and measure the capacity of the team against the work to be performed. If the team does not have the capacity to complete the work, then some of the work needs to be delayed until a future sprint. However, pressure from stakeholders may force the inclusion of the work in the current sprint regardless of team capacity.

Taking on too much work in the sprint is an indicator for potential technical debt insertion. Teams that have too much work to do are more likely to take shortcuts to get the work done and should be carefully watched and reviewed to ensure that they do not insert technical debt, either intentionally or unintentionally, into the system.

Another indication of technical debt during sprint planning is an excessive number of incomplete stories that carryover from one sprint to another. If development teams are routinely not completing tasks within a sprint, they will soon start to feel pressure to get their work done even if it means taking shortcuts and introducing technical debt. Therefore, developers not completing stories is an indicator that technical debt may soon be introduced into the system.

8.3 Active Technical Debt Identification

Active technical debt identification is the process by which technical debt is identified while it is occurring in the system. Unlike proactive technical debt identification, which finds technical debt prior to making the technical compromise, active technical debt identification finds technical debt items as the work is being executed. Active methods seek to prevent the growth of the interest and fees from intentional technical debt and to identify unintentional technical debt as it is being added to the system. Opportunities for active identification of technical debt include peer reviews and the processing of requests for waivers to requirements.

8.3.1 Peer Review

Peer reviews are a technique where one member of the system development team reviews the work of another member of the team, prior to the finalization of the work and in between the major systems reviews. These reviews are designed to help find defects in the process of completing work instead of waiting for the defects to impact the system [15]. Often going into more detail than major system reviews or control boards, peer reviews allow the reviewer to actively identify intentional and unintentional technical debt through a critical analysis of the performed work. As a result of the review, technical compromises should be removed (if they were unintentional) or brought forward to a control board for approval prior to committing to system changes made by the developer.

8.3.2 Requests for Requirement Waivers

A system developer intends to meet all system requirements. However, during the development process, there are times when the system is unable to meet one or more of the requirements. At this point, the developer can submit a request for a waiver of a requirement. A waiver request is a request to accept a part of the system that may not meet the requirements but may still be acceptable "as is" to the stakeholder [16]. Waivers are typically requested after the system is implemented and integrated and are typically discovered during system testing.

As an example, consider a new sports car with a requirement to go from 0 to 60 mph in under six seconds. The design is compliant to the requirement, however, when tested, it takes 6.2 s to reach 60 mph, which does not meet the requirement. The system developer could study the results to determine why it took an extra 0.2 s and could redesign the entire vehicle, which is a time consuming and costly process. On the other hand, they could request a waiver for the 6-s requirement. If granted, the waiver gives relief from the requirement and enables the system development to proceed even though it did not meet its performance goals. The decision on whether or not to grant the requirement waiver request rests with the system stakeholders.

The processing of requests for waivers are times to actively identify technical debt. The conditions that resulted in the system not meeting its requirements are already present and therefore so is the associated technical debt. During the waiver review process, the system developer and the stakeholder need to consider the consequences of accepting the waiver—how does the underperforming system or part of the system impact the rest of the system? How does it impact the user? How important are these impacts? In the above example, accepting the waiver would result in a fee—the 0.2 s longer that it takes to reach 60 mph. If the car is designed for standard everyday use, then this deviation from requirements may not be important. If the car is designed for racing, then it might be critical and failing to meet the requirements could affect the sales and long-term health of the system. Understanding the use of the system and the technical debt associated with the waiver is essential to making the correct decision as to whether or not the waiver request should be accepted.

8.4 Reactive Technical Debt Identification–Finding Unintentional Technical Debt

In the previous two sections, we've discussed ways to identify technical debt before it becomes fully entrenched within the system. Unfortunately, a lot of technical debt is unintentional and enters the system unknown to the system developer. Unintentional technical debt is not necessarily nefarious—it may have been inserted as the natural result of system evolution and schedule pressures. However, any latent technical debt has the potential to create negative long-term impacts on the system. Therefore, we need to develop ways of reactive technical debt identification—finding and handling the technical debt that is already in the system. This section provides methods to adapt existing systems engineering processes to look for existing technical debt.

8.4.1 Sprint Demos and Reviews

In Sect. 8.2.3 we discussed how development teams using sprints may overcommit during a sprint, which can lead to technical debt. Sprint teams may not finish all the work they were assigned, which leads to splash: work carrying over to the following sprint. If a team is under pressure to limit splash, then the quality of the completed work may decrease due to schedule pressure. In sprint cycles, the quality of work is assessed at the end of the sprint through a demonstration and a review [17]. This work is completed and part of the system baseline and therefore any technical debt identification here constitutes a reactive approach–work will need to be done to undo the technical compromise, if necessary.

At the sprint demonstration and review, the sprint team presents their completed work and stakeholders have an opportunity to review it. These reviews are a great time to identify technical fees–the user can review the completed parts of the system and give feedback if it meets their expectations. If the user can interact with the system, they can provide insight as to whether or not it will work within the application domain. This feedback is critical to early identification of potential domain debt, enabling the system developer to correct it before delivering the final release of the product.

Sprint reviews should be tailored to have the sprint team define any compromises or shortcuts that they made during the sprint to identify introduced technical debt. For example, a sprint team may have started work on an interface before the interface control documents were completed and therefore took on technical debt. These decisions should be tracked in reviews and the technical debt added to the register such that its impacts can be mitigated.

Sprint reviews should also examine if the work executed during the sprint took longer than the estimates made by the team. For new sprint teams, incorrect estimations are a part of the team forming process. However, for experienced teams, stories that take longer than expected may be an indicator of technical debt—the developer has to work around earlier technical compromises. Therefore, work with longer than expected durations may be an indicator of unintentional technical debt in other parts of the system.

8.4.2 System Gate Reviews

System gate reviews are the decision points during the system lifecycle that are of sufficient importance to be included in the master program schedule and to have established entry and exit criteria. These reviews ensure that the system is meeting its objectives in all system dimensions and that it is delivering acceptable value to stakeholders. The gate reviews provide opportunities to ensure that the system is meeting its requirements, is deliverable on budget and on schedule, and that there is an acceptable level of risk with proceeding with system development [4].

Gate reviews vary based on the development method used by the system developer. For example, a project developed with a sequential model may have a systems requirement review (SRR), a preliminary design review (PDR), critical design review (CDR), production readiness review (PRR), test readiness review (TRR), and an operations readiness review (ORR). These reviews occur at the completion of the stages of the system lifecycle. A system using an adaptive development method may rely on sprint reviews, program increment reviews, and iterative design reviews that occur on a regular schedule instead of upon completion of a lifecycle stage.

Regardless of the timing of reviews, gate reviews provide a prime opportunity to review the progress of the system and to identify any existing technical debt. At each one of these reviews, the system developer should review the state of the technical debt register as part of the determination of the acceptable level of program risk. High

Table 8.2 Artifacts to review for technical debt during gate reviews

Lifecycle stage	Gate review at stage exit	Artifacts to consider	Technical debt types created
Needs analysis	Alternative systems review (ASR)	CONOPS	Architecture, documentation, domain, mod/sim
Requirements definition	System requirements review (SRR)	SRD/SPS, RVTM	Architecture, documentation, domain, infrastructure, mod/sim, requirements, test
Preliminary design	Preliminary design review (PDR)	Design documents, ICDs, Key performance parameters (KPP)	Design, documentation, domain, mod/sim
Critical design	Critical design review (CDR)	Design documents, ICDs, KPP	Design, documentation, domain, mod/sim, test
Integration	Test readiness review (TRR)	Test plans and procedures	Architecture, documentation, domain, implementation
Verification and validation	System verification review (SVR)	Test reports	Documentation, domain, mod/sim, test
Operations	N/A	User feedback	Documentation, domain

levels of unmitigated technical debt should be treated like high levels of risk in the decision of whether or not to approve the gate review. Gate reviews are opportunities for the entire system development team and stakeholders to come together and review the state of the entire system. Therefore, they are an ideal place for understanding long-term impacts of design decisions that have already been made.

When reviewing artifacts during a gate review, the review team should look for any changes to the originally planned design or predicted performance. Work that is incomplete must be examined and any nonconformance to system requirements identified. The results of trade studies performed during the stage should be scrutinized to determine if technical debt was added to the system. If the review is for a system delivered by a contractor, it is critical to have sufficient insight into the system being developed to ensure there is no technical debt. Table 8.2 summarizes the set of artifacts to consider [18] and types of technical debt to look for at each gate review. Detailed descriptions of each lifecycle phase can be found in Sect. 6.1.

8.4.3 Failure Review Boards

When a formal test fails during system development, the developer will convene a failure review board (FRB) to determine the root cause of the failure. A typical FRB process is shown in Fig. 8.9. The process starts when an anomaly is observed in the

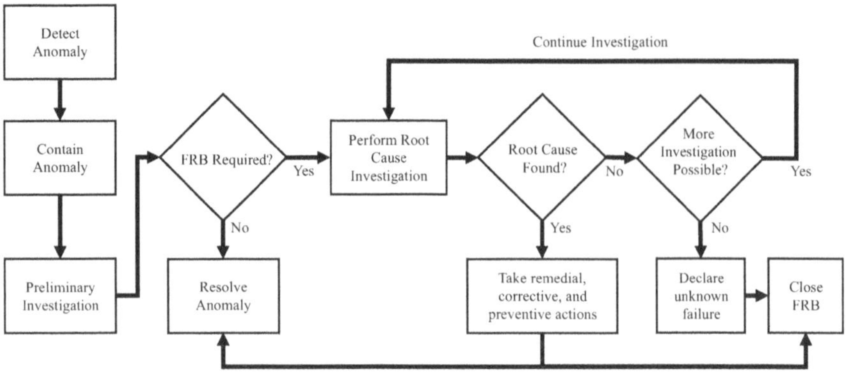

Fig. 8.9 Typical FRB process, adapted from [19]

test. The test configuration is maintained until the anomaly is confirmed as a failure, which kicks off the FRB process. The preliminary investigation phase attempts to determine the cause of the failure and to preserve as much data as possible. At the conclusion of this phase, the system developer decides if an FRB is required. If it is, then the test configuration is broken and the root cause analysis begins. Analysis continues until the root cause is found or the failure is confirmed to be of an unknown cause (due to lack of data to provide definitive evidence). If the root cause is found then the FRB implements corrective actions [19].

Root cause analysis is the process by which the system developer identifies what actually cause the failure. There are numerous tools and techniques to perform root cause analysis, including fault trees, fishbone diagrams (also known as Ishikawa diagrams) [20], and the "five whys method" [21]. These tools and techniques focus on identifying where the part failed and why it failed. For example, electronics on spacecraft may fail due to 'tin whiskers,' which occurs when solder containing tin is used in assembling the components. In zero-gravity environments, the tin can grow thin strands that can cause short circuits [22]. A simplified example of a fishbone diagram leading to the conclusion of tin whiskers on a spacecraft is shown in Fig. 8.10.

These tools and techniques identify the root cause of the failure, in terms of the failed part or process. What they don't always get to is why that part or process was used. From a technical debt perspective, we are concerned with why we took the action that created the root cause of the failure. Why was tin solder used in parts for the spacecraft? Did we know this was a risk and decided to make a technical compromise to save money? Did we not know any better? Technical debt analysis of the failure pushes to drive the root cause back to the technical compromise and short-term benefit that preceded the actions that led to the root cause of the failure. By doing so, we identify the technical debt in the system which can help understand if the issue is a systemic problem or a single point of failure.

Figure 8.11 shows the FRB process modified to include an assessment of technical debt. While multiple tools can be extended to assess the technical debt associated with a failure, we recommend asking the following questions:

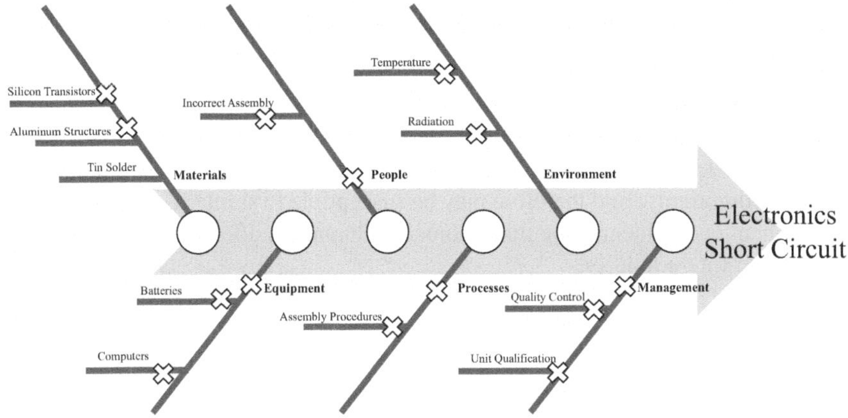

Fig. 8.10 Example fishbone diagram

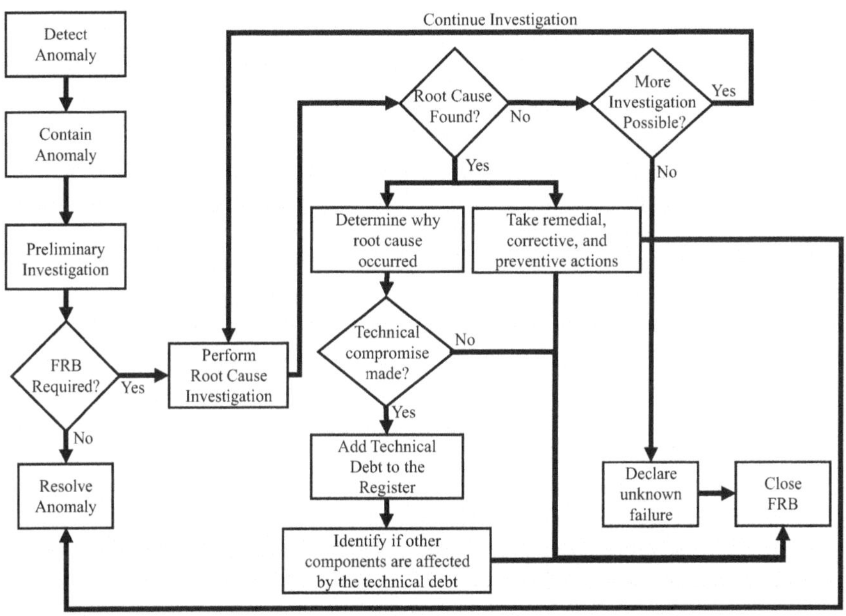

Fig. 8.11 FRB process modified to assess technical debt

- What were the actions taken by the system developer that led to the root cause of the failure?
- Were these actions the result of a technical compromise?
- What was the actual technical compromise made?
- What was the benefit achieved by making the technical compromise?

- Was the technical compromise made throughout the system or only on a single instance?

By answering these questions, the system developer leverages the FRB as a method to identify existing technical debt. With the technical debt identified, artifact traces can be conducted to determine if other components may be affected by the original technical compromise and therefore may be susceptible to similar issues later in the development cycle. Identifying this technical debt and its affected components can help prevent future failures.

8.4.4 Test Results

Not every test fails or requires an FRB. However, that does not mean that every test is completely successful. Test results may show that the system does not perform as expected in the test conditions. The system may still meet its requirements, but test results do not match the predicted behavior. While not every underperformance in test is the result of technical debt (see Sect. 3.5), technical debt should be investigated as a potential source of the underperformance.

Underperforming in a test occurs either because the predicted test performance is inaccurate or because the unit under test does not meet the predicted expectations. Modeling and simulation debt can lead to inaccurate performance predictions. If the system is not modeled to high enough fidelity or in the proper environmental conditions, then the predictions of performance may be incorrect. In this case, the principal would be the extra time required to update the model, the interest is the extra time required to rerun the model and rerun the test, and the taxes are the time it takes to update the test plans and procedures to reflect the new predicted performance. Looking at the models and simulations based on test results can identify this technical debt and enables the system developer to repay the debt before conducting additional tests, if desired.

If the predicted performance is correct, but the unit under test fails to meet the performance, then the system developer should look for design and implementation debt. Was there a compromise made during the design or implementation process that reduced the performance? Remember, a poor design is not technical debt, but a technical compromise made during the design or implementation phase is technical debt. This distinction is important, because a bad design needs to be redesigned, while a technical compromise may be able to be corrected without the need for a redesign. For example, using a part similar to (but not exactly the same as) the final part during the build of a prototype due to part availability represents a technical compromise. If the unit under test does not meet the expected performance due to this part, then the part can be replaced with the actual part (repaying the principal) and the test can be rerun (the interest). This process does not require a system redesign.

Analyzing test results, even when the test is successful provides a reactive mechanism for identifying technical debt. Underperformance can be an indicator of technical debt and should be investigated prior to starting full production of the system.

8.5 Summary

In this chapter, we identified the technical debt timeline and reviewed techniques for proactive, active, and reactive technical debt identification. It is desirable to make technical debt identification proactive to enable early mitigation of interest and fees. However, proactive identification is not always possible within system development. Therefore, active and reactive techniques are necessary to be able to find technical debt within the system and enable proper management of the technical debt.

Reactive techniques are especially important to find unintentional technical debt, which may already be present in the system. By adding a focus on discovering technical debt items to critical review boards and processes, the system developer can identify these latent items to discover the risk to the system from existing technical debt. In the next chapter, we will look at how to measure technical debt items to build understanding of the potential impacts to the system and the cost to remove the technical debt.

References

1. Kleinwaks, H. (2023). *Avoiding Technical Bankruptcy in System Development: A Process to Reduce the Risk of Accumulating Technical Debt*. Doctoral Dissertation, Colorado State University, Department of Systems Engineering.
2. Kruchten, P., Nord, R., & Ozkaya, I. (2019). *Managing Technical Debt: Reducing Friction in Software Development*. Addison-Wesley Professional.
3. Allaverdi, D., & Browning, T. R. (2020). A methodology for identifying flexible design opportunities in large-scale systems. *Systems Engineering, 23*(5), 534–556. https://doi.org/10.1002/sys.21548
4. Walden, D. D., Roedler, G. J., Forsberg, K. J., Hamelin, R. D., & Shortell, T. M. (Eds.). (2015). *Systems Engineering Handbook* (4th ed.). John Wiley and Sons.
5. SEBoK Authors. (2024, November 25). *The Influence of Project Structure and Governance on Systems Engineering and Project Management Relationships*. SEBoK. Retrieved April 18, 2025, from https://sebokwiki.org/w/index.php?title=The_Influence_of_Project_Structure_and_Governance_on_Systems_Engineering_and_Project_Management_Relationships&oldid=72523
6. SEBok Authors. (2024, November 25). *Risk (glossary)*. SEBoK. Retrieved April 18, 2025, from https://sebokwiki.org/w/index.php?title=Risk_(glossary)&oldid=73237
7. Defense Acquisition University. (n.d.). *Risk Management Board and Risk Working Group*. Defense Acquisition University. Retrieved April 18, 2025, from https://www.dau.edu/cop/risk/resources/risk-management-board-and-risk-working-group
8. SEBoK Authors. (2024, November 25). *Fundamentals for Digital Engineering*. SEBoK. Retrieved April 19, 2025, from https://sebokwiki.org/w/index.php?title=Fundamentals_for_Digital_Engineering&oldid=72443

9. United States Department of Defense, Office of the Under Secretary of Defense for Research and Engineering. (2023). *DoD Instruction 5000.97 Digital Engineering.*
10. Delligatti, L. (2014). *SysML Distilled: A Brief guide to the Systems Modeling Language.* Addison-Wesley.
11. Noguchi, R. (2024). The Nature of Technical Debt in the Development of Descriptive Models for MBSE. *INCOSE International Symposium, 34*(1), 516–526.
12. Dassault Systèmes. (2026). *Cameo Systems Modeler*. Dassault Systèmes. Retrieved February 22, 2026, from https://www.3ds.com/products/catia/no-magic/cameo-systems-modeler
13. Montoya, R. (2024, August 27). *Tire Pressure Monitoring Systems (TPMS).* Edmunds. Retrieved May 4, 2025, from https://www.edmunds.com/car-safety/tire-pressure-monitors-can-you-rely-on-them.html
14. United States Department of Energy. (2014, June 23*). Fact #826: June 23, 2014 The Effect of Tire Pressure on Fuel Economy.* United States Department of Energy. Retrieved May 4, 2025, from https://www.energy.gov/eere/vehicles/fact-826-june-23-2014-effect-tire-pressure-fuel-economy
15. Wikipedia contributors. (2023, July 2). *Technical peer review.* Wikipedia, The Free Encyclopedia. Retrieved April 18, 2025, from https://en.wikipedia.org/w/index.php?title=Technical_peer_review&oldid=1162948005
16. Defense Acquisition University. (n.d.). *Waivers.* Defense Acquisition University. Retrieved April 18, 2025, from https://www.dau.edu/glossary/waivers
17. Cohn, M. (2010). *Succeeding with Agile: Software Development Using Scrum.* Addison-Wesley.
18. United States Department of Defense, Office of the Under Secretary of Defense for Research and Engineering. (2021). *Systems Engineering Guidebook.* United States Department of Defense. Retrieved from https://ac.cto.mil/wp-content/uploads/2022/08/Systems-Eng-Guidebook_Feb2022-Cleared.pdf
19. Hecht, T. C., & Air Force Space Command. (2011). *Failure Review Board Guidance Document.* Aerospace Corporation. Retrieved from https://aerospace.org/sites/default/files/maiw/TOR-2011(8591)-19.pdf
20. Wikipedia Contributors. (2025, April 28). *Ishikawa diagram.* Wikipedia, The Free Encyclopedia. Retrieved May 23, 2025, from https://en.wikipedia.org/w/index.php?title=Ishikawa_diagram&oldid=1287785219
21. Wikipedia contributors. (2025, April 7). *Five whys.* Wikipedia, The Free Encyclopedia. Retrieved May 23, 2025, from https://en.wikipedia.org/w/index.php?title=Five_whys&oldid=1284487007
22. Epp Schmidt, T., & Leidecker, H. (2023). *Long-Term Tin Whisker Risk Mitigation using a Conformal Coating.* NASA Electronic Parts & Packaging Electronics Technology Workshop.

Chapter 9
Measuring Technical Debt

Abstract Measurement is the second process in the technical debt management cycle. In this process, technical debt items are quantified by estimating the principal, interest, taxes, and fees. Guidelines for developing a measurement system are provided and algorithms developed to measure principal, interest, and taxes in units of duration. Fees are measured based on the normalized impact to figures of merit. The total system technical debt is then computed across all technical debt items.

Keywords Technical debt measurement · Principal estimation · Interest estimation · Taxes estimation · Fee estimation · System technical debt

The second process in the technical debt management cycle is measurement. The goal of this process, shown in Fig. 9.1, is to quantify each technical debt item. Quantifying the item allows the system developer to determine the potential impact and importance of each item within the greater system context.

However, measuring technical debt items is one of the largest challenges in managing technical debt. By its very nature, technical debt represents uncertain future impacts which complicate the measurement. Yet measuring technical debt is necessary to enable prioritization of repayment efforts and to control the spread of the technical debt throughout the system. We must be able to quantify technical debt to determine the proper actions and resources required to remove it. Fig. 9.2, taken from the technical debt context map, shows the quantitative parameters of a technical debt item.

To measure technical debt, we will look at the four quantitative components: principal, taxes, interest, and fees. Computing each of components is complicated and the published research has yet to provide simple methods that connect all four components. Methods to estimate the principal in software engineering include determining the cost of the rework within a release [1], real options analysis [2], and other financial analogs [3]. Some authors suggest only estimating the principal and interest as high, medium, and low until further refinement is required [4]. Other work has provided various complicated methods for estimating interest, but these tend to be specific to problems or software systems and not easily generalizable.

H. Kleinwaks, *Managing Technical Debt in Systems Engineering*,
https://doi.org/10.1007/978-3-032-30360-8_9

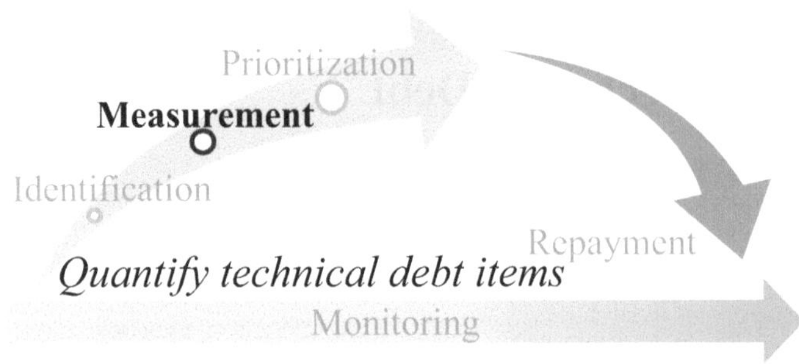

Fig. 9.1 Measurement process

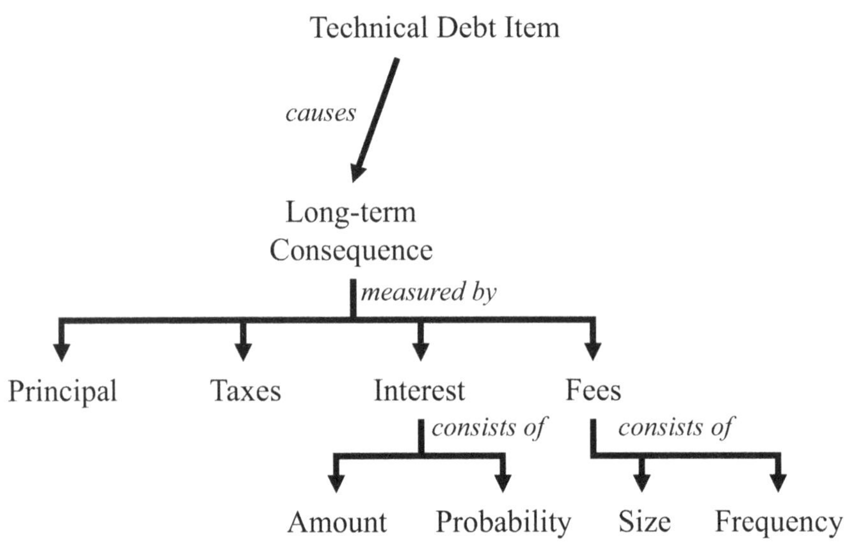

Fig. 9.2 Quantitative components of a technical debt item

In this chapter, we will develop a simplified model for estimating principal, interest, and fees that can be broadly applicable across different projects. The goal of this model is to enable the system developer to measure each technical debt item on a consistent scale. The developer can then compare the technical debt items and prioritize their repayments, based on the effort to repay and the potential future impacts.

To create our technical debt measurements, we will use the following guidelines:

1. Measurements must produce quantifiable results
2. Measurement techniques must be independent from the system being analyzed
3. Measurements must account for the uncertainties associated with technical debt
4. Measurements must be comparable across different technical debt items such that items with more impactful technical debt produce larger measurements

Adhering to these guidelines ensures that the measurement framework we create, and therefore the technical debt measurements produced, can be used to quantitatively compare different technical debt items from multiple disciplines and projects. Establishing this common basis of measurement then enables the measurements to be considered as an unbiased contributor to the prioritization of technical debt items. These measurements are not the only factors considered in technical debt prioritization (see Chap. 10), but they are important ones to consider. Measured technical debt items can also be used to assess the likelihood of technical bankruptcy, which we will discuss in Chap. 12.

9.1 Estimating Principal

Principal is the measurement of the compromise made in the performance dimension to achieve a short-term benefit. The measurement of the compromise is analogous to the amount borrowed in financial terms: the measurement is what we are taking from the system to achieve our desired benefit. In finance, you borrow $40,000 from the bank (principal) to buy a car (benefit). When creating technical debt in systems development, you borrow from the system performance (principal) to accelerate schedule or save budget (benefit).

Measuring the reduction in performance is complicated. In some instances, the reduction in performance can be clear and quantitatively stated. For example, using aluminum instead of steel in structures reduces the amount of weight that the structure can support. Similarly, using a less powerful computer increases the runtime of a software program. While these measures are intuitive and quantifiable, it is challenging to compare them to determine which item has more technical debt. How do you correlate system runtime to system strength? Therefore, the reduction in raw performance metrics, while absolutely necessary to track as part of the technical debt item, does not make for a comparable measurement of principal, violating guideline #4.

Instead of measuring principal by the reduction in performance we can measure it by what is saved. When we borrow money to buy a car, we are not just acquiring the money, we are also saving the time it would take to acquire the money through other means. If you save $500 each week then it would take you 80 weeks to reach the $40,000 required to buy the car (assuming the car price hasn't gone up in that time frame as well!). By borrowing the money from the bank, you get the money today and lock in the price of the car. You are not just borrowing the money; you are also saving time. Can this idea be applied to the calculation of the technical debt principal?

The causes of technical debt include schedule pressure, cost pressure, technical trades, and political pressure. When schedule pressure is applied, the system developer makes a technical compromise to save schedule. Therefore, the principal can be calculated based on the amount of time saved. If not executing tests saves us 10 days

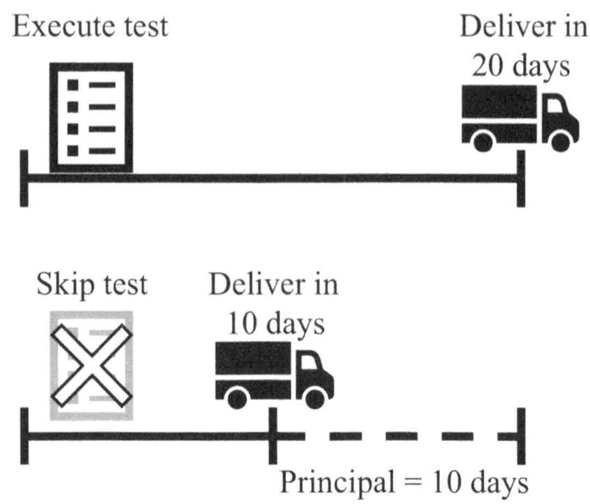

Fig. 9.3 Principal measured in duration

of testing and enables an earlier release, then the principal is 10 days. This comparison, shown in Fig. 9.3, is fairly straightforward–we save the time it would take to have implemented the original solution and that time saving constitutes the principal. This measurement meets guidelines 1 (it is quantifiable) and 2 (time saved can be calculated for any system). Guideline 3 can be satisfied if the 10 days of savings is modeled with uncertainty: there is a probability distribution around the 10 days that can be modeled and the associated interest will also be probabilistic. Guideline 4 is the challenging one. While time-savings can be computed for any type of technical debt resulting from schedule pressure, can it also be computed for technical debt that results due to other causes?

To answer this question, let's examine technical debt that arises from cost pressure. Cost pressure forces the system developer to make technical compromises to reduce the cost of the system. This principal could easily be measured by the amount of money saved, similar to the discussion of time saved due to schedule pressure. But can we convert the money saved to time saved, or should we do it the other way around?

In general, it is relatively easy to go from time to money–you can take the employee salary and multiply it by the number of work hours and get a cost value. Costs for delays can be estimated and costs of equipment and running tests calculated. Going the other way is not as simple–how do you convert the difference in price between steel and aluminum into a temporal value? On the contrary, it may be easier for system developers to think in terms of time. A test engineer can easily tell you how long a test will take to run but may not have access to all the information required to tell you how much the test will cost.

Which should we use? Again, we need to examine what we are trying to measure with technical debt. The principal is the measurement of the compromise in the performance dimension and therefore can only be repaid through effort expended in the performance dimension. This effort may include redesigning the system, rerunning tests, refactoring code, or other activities. These activities can be measured in the time it takes to execute them, even if the primary benefit is a cost savings. Let's look at the example of replacing steel with aluminum in a car because the aluminum is cheaper. The reduction in performance is a decrease in the safety factor for the car, due to the reduction in strength of the car's frame. Theoretically, the car could be redesigned to achieve the same safety performance with an all-aluminum frame, and therefore the true technical compromise is replacing steel with aluminum without performing the redesign work. We can measure the effort, and the time, to perform the redesign work, and also the impact on any future design work due to the presence of the aluminum frame. We can relate cost to time, as long as we really think about the compromises being made.

Given this approach, we will use time as the measurement unit for principal. Let's apply it to the two other primary causes of technical debt: technical trades and political pressure. Technical trades involve making a technical compromise to achieve a benefit in another performance dimension. We can use the same car frame example here. Consider that a car likely has multiple figures of merit, such as safety and gas mileage. Switching from steel to aluminum produces a lighter car and increases the gas mileage but reduces the safety. If the budget and schedule are not a concern, then this switch would be a technical trade: compromising one aspect of the performance dimension to achieve a benefit in another aspect. The principal would be the time it takes to design a car that achieves the specified performance in both figures of merit.

Political pressure forces technical compromises to achieve benefits in one of the system dimensions. Therefore, the same analysis applies to measuring the principal of political pressure as it does for the other three sources and the principal associated with technical debt accrued due to political pressure can be defined through a temporal measurement.

With all the above arguments, we can finally state how to measure principal:

> Principal is measured as the development time required to restore the system performance that was reduced through the technical compromise.

This approach conforms with all four guidelines: duration estimates are quantifiable, the measurement is independent from the system–all systems can estimate development timelines, duration estimates can be modified to account for uncertainties, and duration estimates can be compared across multiple technical debt items through consistent unit applications.

The calculation of principal will be dependent upon the unique combination of the technical debt item and the system under development. Equation (9.1) defines the principal as the time to restore the system to its desired performance level. Note that the principal is only calculated based on the state of the system at the time the technical compromise was made and the technical debt was incurred. Work to restore components added after the compromise was made is included as taxes. The subscript 0 is used to indicate that the principal is the baseline restoration cost.

$$P = t_{\text{restore},0} \tag{9.1}$$

Table 9.1 Examples of measuring principal

Technical debt statement	Principal
To meet the delivery schedule, the decision to perform a limited amount of thermal vacuum testing on a satellite was made, resulting in the potential for unknown performance at high payload duty cycles	2 weeks of time to run the full suite of tests
Unintentionally, a decision to use an older version of the computer operating system was made, resulting in the potential for increased downtime due to more frequent patches	2 months to upgrade the operating system to the current version and test the code
To reduce the model execution time, the decision to use low-fidelity models instead of optimizing the model code was made, resulting in the potential for not meeting performance requirements	3 months to optimize the model code

Table 9.1 identifies several examples of technical debt and the associated principal.

9.2 Estimating Interest

Our choice of unit for measuring principal implies that we should use the same unit for measuring interest. Therefore, we consider interest by looking at the increase in the duration of system development due to the presence of technical debt. This unit choice is consistent with the definition of technical debt interest: the expected value of additional units incurred by the system developers in the performance dimension due to the presence of unpaid principal. The interest, like the principal, is measured based on the duration of the task due to the unpaid principal. This unpaid principal adds scope to each subsequent task, increasing its duration.

Interest is a probabilistic quantity, divided into the interest amount and the interest probability. To estimate the interest, we need to determine both the amount and probability. Each affected component will have its own interest amount and interest probability, and therefore the expected value of the total interest is accumulated across each affected component, as shown in Eq. (9.2). Note that the total interest on each component could be calculated through probability distributions instead of a single probability multiplied by the interest amount.

$$\bar{I} = \sum_{i=0}^{n} a_i r_i \tag{9.2}$$

9.2.1 Estimating Interest Probability

Estimating the interest requires first identifying all of the components affected by the unpaid principal. How does a component get affected by unpaid principal? First, the component must interface with the debt-holding component. These interfaces can be identified by analyzing the system documentation for the artifact associated with the technical debt item. Any interfaces with that artifact should be traced throughout the system design. Any component that occurs on that trace could see interest from the affected artifact. MBSE models can assist in performing these traces, as discussed in Sect. 8.2.2.

To understand the estimation of interest, let's work through an example. Figure 9.4 shows the design of a system where component 2 provides the primary interface to component 1. According to the ICD, component 2 is supposed to have two connections to component 1, a USB-A connection and an Ethernet connection. The developers of component 2 looked at the information flow across the interfaces and determined that they did not need the data that came across the USB-A interface and therefore did not implement it to save time and money. However, when component 3 is added to the system, that data is now required, forcing the system developers to either reestablish the connection between components 1 and 2 (repay the principal) or to take extra time to figure out how to connect components 1 and 3 (the interest). Similarly, when component 4 is added, another workaround needs to be implemented, costing more time and money (more interest). The list of affected components can grow quite quickly, which is why it is critical to manage technical debt!

Once we have the list of potentially affected components, the next step is to determine the interest probability for each of the components. Doing so requires an understanding of the areas within the artifact that are affected by the technical debt. Any component that directly interfaces with these artifacts has a high probability of being impacted by the technical debt. Any component that follows along the design flow, but does not directly interface with the affected artifact also has a probability of being impacted by the technical debt, but that probability should be lower–the primary interface can account for some of the changes, but if it needs to change then those changes can cascade throughout the system. Components that interface with the artifact but do not use the affected interface also have a lower probability of interest, but that interest is not zero: if the artifact needs to be updated, then the other interfaces may change as a result.

The interest probability for each component is the likelihood that the interest amount for that component will need to be paid by the system developer. The probability can be estimated with multiple methods and should be based on the experience of the system developer and the unique technical compromise. Some compromises result in more certain interest payments: for example, deciding not to procure updated test infrastructure is almost certain to increase the duration of testing while shortening user testing may not impact the acceptance of the final product. Equation (9.3) defines the interest probability for a specific component:

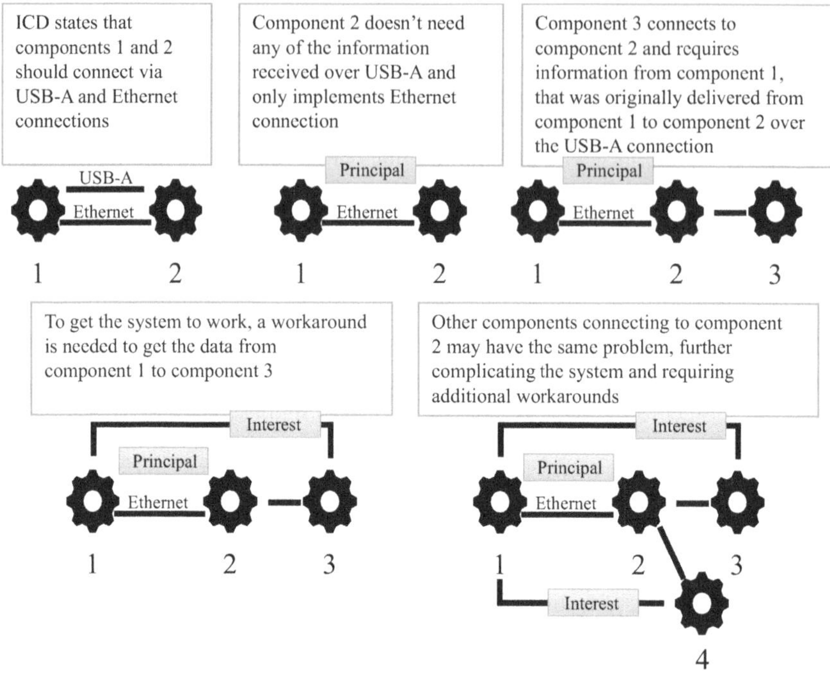

Fig. 9.4 Example of interest build up due to incomplete ICD implementation

$$r_i = p(\text{interest amount occurs})_i \tag{9.3}$$

In Fig. 9.4, the interest probability would be determined by assessing the likelihood that future components need to access the data provided across the USB-A interface, which was removed.

9.2.2 *Estimating Interest Amount*

Similar to the interest probability, the interest amount for each component needs to be estimated. The interest amount is a function of the state of the system, which means that the number of affected components and the state of their development affect the interest amount. Interest must be in the same units as the principal and the interest probability is unitless. Therefore, the interest amount must be measured as a duration parameter and is evaluated by estimating the duration required to complete the additional work due to the unpaid principal.

To establish the interest amount, we need to determine the additional effort required to work around the technical compromise. There are two aspects that need to be estimated in order to determine the interest amount: the nominal amount of work required and the additional amount of work required due to the presence of technical debt.

Work is required to add a component, regardless of the presence of technical debt. For example, there may be technical debt in a system because a component was not fully tested. A new component is added to the system, and the work required to add the component must occur whether or not the initial system test was complete. This is the baseline work. Changes to the new component due to design flaws of the original system that were not revealed because of the limited testing are interest on the technical debt–the flaws in the original system are unknown and may or may not occur. Therefore, the interest amount is mathematically defined as the time to work around the technical debt, as shown in Eq. (9.4):

$$a_i = t_{\text{workaround},i} \tag{9.4}$$

For each affected artifact, the system developer estimates the amount of additional scope required due to the presence of the unpaid principal. In the example shown in Fig. 9.4, the interest amount would be the estimate of the work it takes to directly connect components 3 and 4 to component 1, instead of relying on the planned connection to component 2.

9.3 Estimating Taxes

The presence of technical debt impacts the addition of new system components both when the component is added and when the technical debt is repaid. We defined the former as the interest and the latter as the taxes on the system. Unlike interest, taxes are not probabilistic. Like principal, taxes represent the known quantity of work required to rework a component to restore the system performance. Taxes effectively increase the principal–for every component added to the system that is affected by the technical debt, there is an additional cost to restore the performance of the system.

Therefore, taxes can only be measured after a component has experienced the effect of the technical debt interest. If the system developer had to implement a work-around due to the presence of technical debt, or implemented a solution that relied on a previous work-around, then that component is subject to taxes and there will be additional work required to remove the work-around once the principal is repaid.

Since taxes increase the principal, taxes must be measured using the same methods as principal. Taxes are computed by determining the development time required to rework the component once the principal has been repaid. Taxes do not necessarily equate to the interest amount for a component, as the work does not necessarily involve removing the workaround that was applied if the interest had to be paid. Sometimes, the workaround applied when the interest was paid does not need to be modified, and sometimes it will require significant effort. Equation (9.5) defines the taxes:

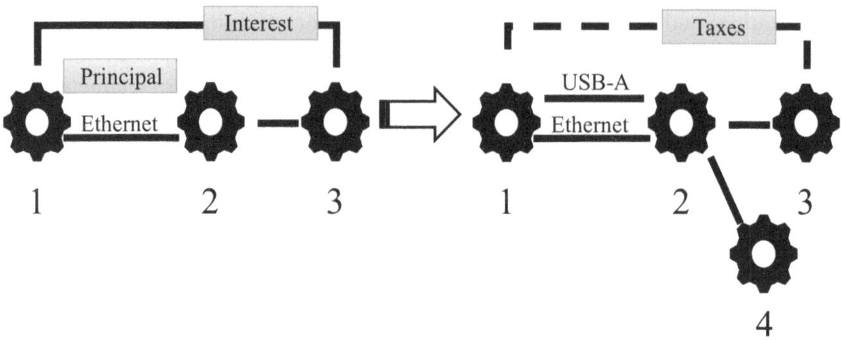

Fig. 9.5 Example of taxes and principal repayment

$$T_i = t_{\text{restore},i} \tag{9.5}$$

To understand how taxes come about, let's return to Fig. 9.4. Assume that prior to adding component 4, the system developer decides to reestablish the interface between components 1 and 2, instead of trying to do workarounds for multiple components. The system developer then has to perform additional work to remove the direct connection between components 1 and 3 and to create the proper data flows between components 2 and 3. This work is the taxes: the cost to remove the workarounds that were introduced when paying the interest. The work to reset the connection between components 1 and 2 is repaying the principal, as that was the original technical compromise. This dynamic is shown in Fig. 9.5, which shows the removal of the principal and the payment of taxes, which allows component 4 to be added without any additional interest payments.

9.4 Estimating Fees

Fee-bearing technical debt requires different units. The fees paid by the user exhibit a decrease in the performance of the system. These units are necessarily variable, based on the specific performance parameter. The amount borrowed may be either in the time, cost, or performance dimension. For example, if a car manufacturer uses aluminum instead of steel in the car frame, the overall safety of the car will be reduced. The manufacturer may choose to make this change since aluminum is cheaper than steel (short-term benefit of cost savings), and the car may still meet safety ratings. However, the achieved safety ratings may be lower than desired. Therefore, the change in safety rating is the change in performance. The fee paid by the user is an increased risk of injury every time they drive the car.

Fee-bearing debt is inherent in the system and cannot be paid back without another release of the system. The fees occur when the system is used and can be measured by examining the figures of merit of the system. A fee represents a reduction in a figure of merit, such as usability, safety rating, or miles per gallon. In the above example, the increased risk of injury represents a reduction in a figure of merit.

However, measuring fees using figures of merit does not appear to comply with the four measurement guidelines. Figures of merit are unique to a specific system and it is not easy to compare the impact of a figure of merit reduction across multiple figures of merit. Additionally, figures of merit are not easily translatable to duration estimates, so we cannot compare fee-bearing technical debt to interest-bearing technical debt. So how can we measure fee-bearing technical debt?

The answer is to split the measurements of fee-bearing debt into the fees and the principal and taxes. The principal and taxes are measured using the same mechanism as interest bearing debt. Calculating principal and taxes in the same way for fee-bearing and interest-bearing debt will allow for a comparison of these values across all technical debt items.

Instead of calculating the interest as a duration value, fee-bearing debt must consider the reduction in performance experienced by the user of the system. To measure this reduction, we calculate technical fees based on reductions in the system's figures of merit. The fee amount, g, is this reduction in the figure of merit, normalized by the figure of merit value, as shown in Eq. (9.6). By normalizing the reduction in the figure of merit, we can compare different technical fees according to their relative impact on performance.

$$g = \frac{\Delta\text{FOM}}{\text{FOM}} \tag{9.6}$$

The fee frequency measures how often a user has to pay a technical fee and accounts for the probability of the user experiencing the reduction in performance with each interaction. The technical fees for a technical debt item are calculated as the product of the normalized fee amount, g, the fee frequency, f, and the elapsed time, t, as shown in Eq. (9.7). While the fee amount should be a static number, the free frequency could be calculated as either a fixed rate or from a probability distribution.

$$F = g * f * t \tag{9.7}$$

With these definitions in place, we can reexamine the measurement guidelines. Measuring fees based on the balance and the reduction in performance produces quantifiable results (guideline 1). By normalizing the figures of merit, the reduction in performance is independent from the system being analyzed (guideline 2). The uncertainties of the fee occurring are captured via the fee frequency (guideline 3). Finally, the balance of fee-bearing technical debt items can be compared to the balance of interest-bearing technical debt items, allowing comparison between the two types (guideline 4).

As an example, consider a technical debt item that increases the need for system maintenance. This consequence can be measured by determining the system mean time before failure (MTBF). The system will not break every time that it is used, but it will break more frequently than it should. The fee amount would be the repair time–how long the system is unusable when it breaks. The fee frequency can be calculated from the MTBF by estimating how often the system will break. Fee-bearing technical debt contributes to both the technical debt balance (through the associated principal and taxes) and the total fee calculation, which we will review in the next section.

9.5 Calculating System Technical Debt

The system technical debt is the total technical debt within the system and consists of two parameters: the balance and the total cost. The balance is the quantitative cost to the system developer to repay the technical debt and includes the principal and taxes of interest-bearing and fee-bearing technical debt items. The total cost is a measure of the complete impact of the technical debt and includes the interest in addition to the taxes. The total cost of repaying a technical debt item will vary over time as the taxes and interest change. As interest is paid, taxes may be incurred to repair the workaround introduced when paying the interest. Finally, the total fees also need to be calculated to determine the impact to the user.

The balance is calculated as the summation of the principal and taxes of all technical debt items (including interest-bearing and fee-bearing items), as shown in Eq. (9.8). The balance needs to be recalculated as the system evolves, since the value of the taxes is a function of the state of the system: the more the interest that has been paid by implementing workarounds, the higher the taxes required to repair the workaround. Therefore, the balance is also a function of the state of the system, indicated by the subscript t.

$$B_t = \sum_{j=0}^{n} \left(P_j + \sum_{i=0}^{m} T_{t,j,i} \right) \tag{9.8}$$

The total cost is the summation of the balance and the potential interest. The total cost is an expected value due to the probabilistic nature of the interest. The total cost is also dependent upon the state of the system since the balance and interest are both functions of the state of the system. The total cost is calculated using Eq. (9.9). Note that for the fee-bearing technical debt items, the interest terms are zero.

$$\overline{C_t} = \sum_{\mathrm{j}=0}^{n} \left(P_{\mathrm{j}} + \sum_{\mathrm{i}=0}^{\mathrm{m}} T_{t,\mathrm{j},i} + \sum_{i=0}^{n} a_{t,j,i} * r_{t,j,i} \right) \tag{9.9}$$

Figure 9.6 shows the accumulation of principal, interest, and taxes during the lifespan of a technical debt item. The initial technical debt item results in five units of principal. To remove the technical debt item, this amount will need to be repaid. The technical compromise initially is determined to only affect one additional component, whose interest value is two. The interest has not yet been realized, so it has not been paid. Similarly, when another component is determined to be affected by the compromise, its interest is logged but has not been repaid, leaving the balance and amount paid the same, as shown in the table at the bottom of Fig. 9.6.

The total cost, which represents the total amount that might be paid, increases with the addition of the interest-bearing components. In the last column of Fig. 9.6, some of the interest has been repaid. Interest is paid when the associated development efforts take longer than originally planned. Paying the interest involves the implementation of a workaround, which must then be removed when the principal is repaid. Therefore, taxes have been introduced into the system, shown in white. The introduction of taxes increases the balance (the sum of the principal and the taxes). The total cost is updated to reflect both the current state of the interest and the taxes in the system. The total cost is an important parameter as it reflects the total exposure of the system to technical debt if the debt is not repaid. The interest terms account for the time spent executing workarounds and the balance terms account for the time spent correcting the problem.

Finally, we must also calculate the total fee, F, in the system. The fees are the sum of the amount paid per fee occurrence, g, times the free frequency, f, times the

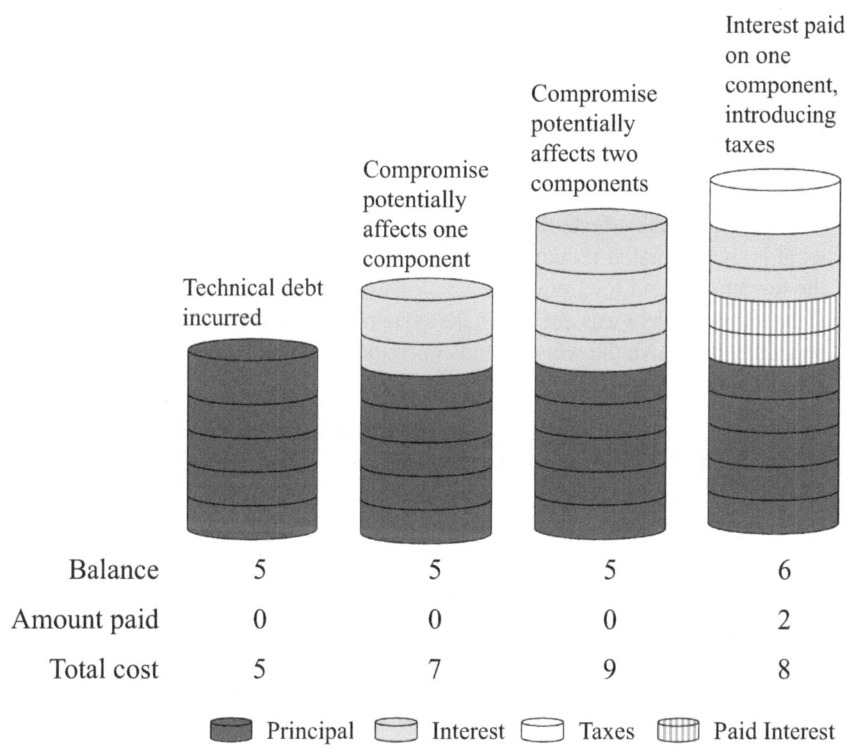

Fig. 9.6 Accumulation of principal, interest, and taxes

elapsed time, t as shown in Eq. (9.10). The total fee is a function of time, since fees paid per technical debt item increase with time.

$$F_t = \sum_{i=0}^{n} g_i * f_i * t \tag{9.10}$$

9.6 Algorithm for Estimating Principal, Interest, Taxes, and Fees

Now that we've established how to estimate principal, interest, taxes, and fees, we can develop an algorithm for quantifying the technical debt in a system. Let's start with a simple set of questions and then develop a detailed flow diagram. Any time the system developer makes a technical compromise, they should ask themselves the following questions to generate a quick measurement of the technical debt:

1. What artifacts are negatively affected by the compromise? This question helps to frame the answers to the following questions.
2. How long would it take to implement the system as designed or intended without this compromise? The work not being done is the principal.
3. Will the compromise affect the system developers or the system users? This question identifies if the technical debt is interest-bearing or fee-bearing.
4. If the compromise affects the system developers, what workarounds are anticipated and how long will they take to implement? This question estimates the interest amount and interest probability.
5. If the compromise affects the system users, what figure of merit is affected and what is the estimated reduction in that figure of merit? This question estimates the fee amount and fee frequency.
6. If the technical debt was already in the system before it was discovered, how long will take to remove the workarounds once the principal is repaid? This question estimates the taxes.

These questions help frame the discussion, and basic estimates can be used to quickly measure the technical debt item. Detailed estimates can be performed as needed and according to project-specific guidelines.

The flow diagram for this process is shown in Fig. 9.7. The same sequence of steps applies to every identified technical debt item and can be rerun when a new technical debt item is identified. The first two steps are part of the definition of the technical debt item: identifying the affected artifacts and the technical compromise. These steps are essential to estimate the potential impacts of the technical debt throughout the rest of the system: the artifact can be traced through the rest of the system components and the compromise indicates changes which were made.

After identifying the compromise, the principal can be estimated. Step 3 is to identify the work required to restore the system performance due to the technical

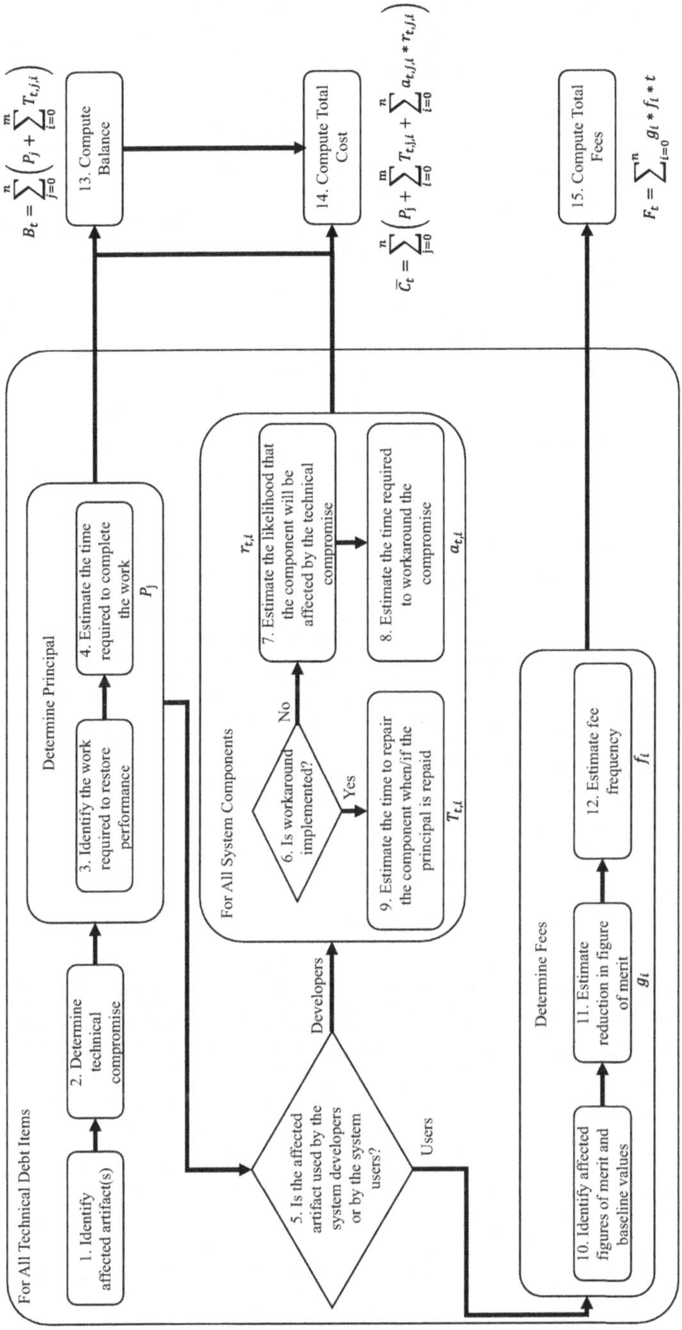

Fig. 9.7 Algorithm for quantifying technical debt in the system

compromise. Step 3 is a necessary prerequisite to step 4, where the actual principal is estimated as the time it takes to perform the work identified in step 3. The identification of the work required in step 3 should be recorded in the technical debt register to use when the time comes to repay the principal.

With the principal estimated, the next step is to determine whether the technical debt is interest bearing or fee bearing. This determination can be made by answering the question in step 5 of Fig. 9.7: "Is the affected artifact used by system developers or by the system users?" Interest-bearing debt affects the system developers and fee-bearing debt affects the system users. Tracing the affected artifact(s) allows this question to be answered. As we have seen, some technical debt may be both interest bearing and fee bearing.

If the affected artifact is interest bearing, then we need to assess all system components for potential impacts from that affected artifact. In step 6 we determine if a workaround for the technical compromise has already been implemented. This question is a critical step, especially when technical debt is identified after it has already been in the system. If the workaround has already been implemented in a component, then that component has already experienced the interest from this technical debt item. Interest includes items such as longer tests and longer development cycles, performing additional work due to missing interfaces or information, or redesign and rework efforts. If a workaround has not already been implemented in the component, then we need to estimate the likelihood that a workaround will be needed (the interest probability, step 7) and the time it will take to implement the workaround (the interest amount, step 8). If a workaround has already been implemented in the component, then we need to calculate the taxes for that component (step 9)–the estimate of the time to remove the workaround once the principal is fixed. A component may see both interest and taxes, depending on its state of development.

If the affected artifact is fee-bearing, then we move to step 10, where we identify the affected figures of merit and their baseline values. Then we estimate the impact to those figures of merit (step 11) and the frequency at which those impacts will be observed (step 12) to calculate the total fees in the system.

Once all technical debt items have been analyzed, we calculate the balance, total cost, and total fees (steps 13–15) such that we have quantified measurements of the system technical debt. Proper technical debt management requires repeating this process every time new technical debt items are identified or existing technical debt items are repaid. The results can then be used to assess how close the system is to technical bankruptcy, which will be discussed in Chap. 12.

9.7 Summary

Measuring technical debt is one of the greater challenges in technical debt management, both in software and systems engineering. The processes presented in this chapter provide a simplified approach to quantifying each technical debt item such

Table 9.2 Summary of technical debt measurements

Component	Measurement
Principal	Development time required to restore the system performance that was reduced through the technical compromise
Interest amount	Additional development time required to work around the unpaid principal
Interest probability	Likelihood that the developer will need to work around the unpaid principal
Taxes	Additional development time to remove the workaround when restoring system performance
Fee amount	Normalized reduction in a figure of merit seen by the user due to unpaid principal
Fee frequency	Rate at which the user sees the fee amount

that they can be prioritized, which will be discussed in the next chapter. The algorithms provided include methods for measuring each technical debt item and for including the impacts of technical debt in schedule risk analysis to account for the impacts of technical debt on the ability to deliver the project on time. Table 9.2 summarizes the measurement definitions for each measurable component of a technical debt item.

The processes used for measuring each technical debt item will necessarily be tailored for each project. System developers are encouraged to develop their own algorithms that can be easily executed to produce consistent and usable measurements. These measurements can be used to inform system stakeholders of the technical debt balance, and therefore the risk of technical bankruptcy (see Chap. 12). Tracking the measurement of each technical debt item and comparing it to the cost to repay the item enables the calculation of the return on investment if the item is repaid. In the next chapter, we'll discuss prioritizing and repaying technical debt items.

References

1. Nord, R. L., Ozkaya, I., Kruchten, P., & Gonzalez-Rojas, M. (2012). In Search of a Metric for Managing Architectural Technical Debt. *2012 Joint Working IEEE/IFIP Conference on Software Architecture and European Conference on Software Architecture*, 91–100. https://doi.org/10.1109/WICSA-ECSA.212.17
2. Abad, Z., & Ruhe, G. (2015). Using Real Options to Manage Technical Debt in Requirements Engineering. *2015 IEEE 23rd International Requirements Engineering Conference (RE)*, 230–235.
3. Ampatzoglou, A., Ampatzoglou, A., Chatzigeorgiou, A., & Avgeriou, P. (2015). The financial aspect of managing technical debt: A systematic literature review. *Information and Software Technology, 64*, 52–73. https://doi.org/10.1016/j.infsof.2015.04.001
4. Seaman, C., & Guo, Y. (2011). Measuring and Monitoring Technical Debt. *Advances in Computers, 82*, 25–46. https://doi.org/10.1016/B978-0-12-385512-1.00002-5

Chapter 10
Prioritizing and Repaying Technical Debt

Abstract Prioritization is the process by which technical debt items are selected for repayment. While the exact method for prioritization will be unique to each project, a prioritization scheme based on the Eisenhower matrix is presented. Repayment is the process where work is performed to remove the technical debt item, either by paying the interest or repaying the principal and taxes. When repaying the principal, the original technical compromise is removed. Multiple repayment strategies are provided, including direct repayment, mitigation, monitoring, and watching.

Keywords Technical debt prioritization · Technical debt repayment

After technical debt items have been identified and measured, the next steps in the management cycle are to prioritize and repay the technical debt items. Prioritization is the act of sorting the technical debt items on the technical debt register to determine an order for repayment. Repayment is the act of doing the work to mitigate the impacts of the technical debt items by paying down the principal, interest, and taxes. This chapter discusses these two related processes.

10.1 Prioritization

The third step in the technical debt management cycle is prioritization, shown in Fig. 10.1. The goal of this process is to obtain a ranked list of the technical debt items to aid the system developer in deciding when to repay each item.

Prioritization starts with the technical debt register developed in the previous two processes. This register lists all technical debt items and their measurements. To prioritize technical debt items, we can draw an analog to the risk management process. In risk management, risks are typically associated with values between 1 and 5 for both likelihood and consequence. Risks can then be depicted on a matrix, as shown in Fig. 7.4. The risk impact is calculated by multiplying the likelihood and consequence score, allowing for risks to be ranked by impact and then prioritized. This process

H. Kleinwaks, *Managing Technical Debt in Systems Engineering*,
https://doi.org/10.1007/978-3-032-30360-8_10

Fig. 10.1 Prioritization process

enables schedule risks to be directly compared to cost risks and performance risks. An alternative method for ranking risks is to monetize the risk consequence and then rank risks by the product of risk likelihood and the monetized cost.

When looking at the set of technical debt items in a system, we can create a plot of the technical debt balance (principal and taxes) against the interest and fees and then divide the plot into quadrants as shown in Fig. 10.2. Interest and fees are plotted on different vertical axes due to their different measurement scales, with interest-bearing technical debt items depicted with solid circles and fee-bearing technical debt items depicted with open circles.

Technical debt items with a high balance require a large cost or effort to correct the debt and therefore must exhibit a corresponding return on investment to the developer. Technical debt items with high interest or fees require the developer or user to work around the debt and have the potential for increased delays within system development and increased risk of poor system performance. We can use this plot to prioritize the debt by combining it with the Eisenhower matrix [1], resulting in Fig. 10.3. The Eisenhower matrix defines tasks based on two categories: their importance and their urgency. Important tasks have high value. Urgent tasks must be performed quickly. Therefore, important and urgent tasks must be done immediately and tasks which are neither important nor urgent may not need to be done at all.

In constructing this matrix, we label the technical debt items with high balances as important and the technical debt items with high interest and fees as urgent. High balance items are important because they require larger durations to correct and must be scheduled. The potential for additional taxes when correcting those high balance items is higher and therefore the correction approach must be more carefully considered. High interest and high fee items are urgent because these impacts cause cascading problems to the system developer through development delays or to the system user through reduced performance. It is urgent to correct these problems because they can prevent the successful release of the system.

Using the Eisenhower approach allows us to bin technical debt items into four categories and to develop a repayment approach for each category:

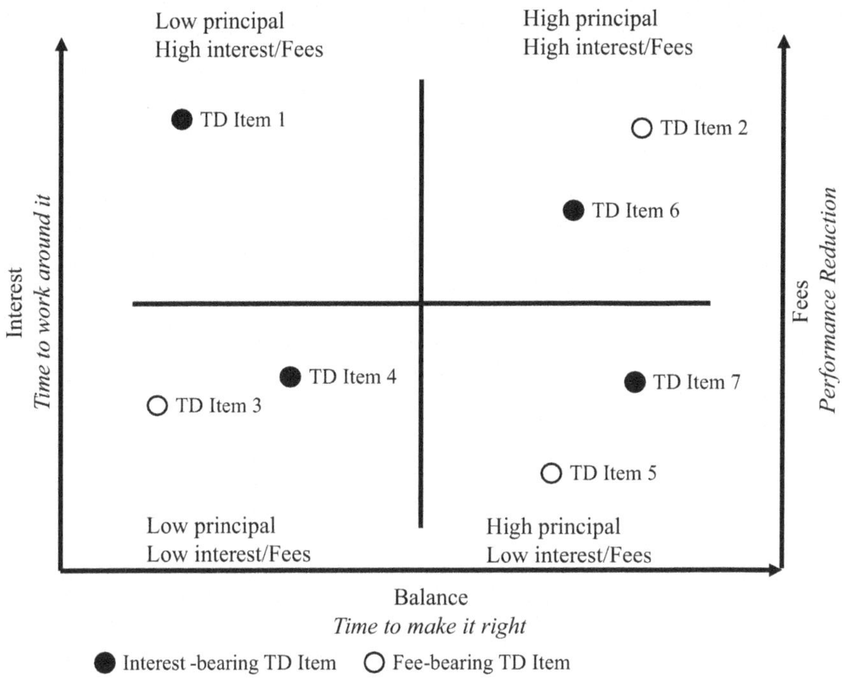

Fig. 10.2 Principal versus interest plot

- Urgent and Important: high balance and high interest/fee items that should be ***mitigated*** by reducing the interest and paying down the principal and taxes
- Urgent and Not Important: high interest/fee items that should have the balance ***repaid*** such that the interest and fees do not continue to grow
- Not Urgent and Important: high balance items that need to be ***monitored*** to ensure that interest and fees do not start to develop, but may not need to be repaid due to the high cost of the principal
- Not Urgent and Not Important: low balance and low interest/fee items that should be ***watched*** but not actively mitigated.

Each of these repayment approaches will be discussed in Sect. 10.2.1.

This matrix approach provides an easy visual for understanding the state of the technical debt within the system. However, it is ultimately up to the system developer to decide on the proper prioritization scheme for their system, factoring in the development schedule, actual measurements for each technical debt item, and availability of system resources. Prioritization should be reviewed at the technical debt review board, discussed in Chap. 11.

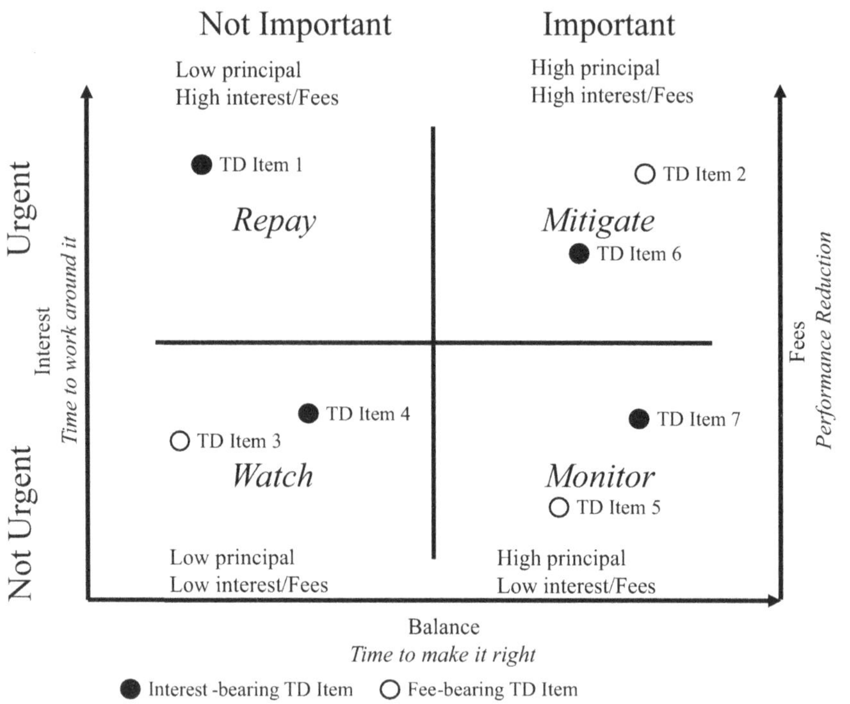

Fig. 10.3 Eisenhower matrix prioritization of technical debt

10.2 Repayment

The next process in the technical debt management cycle is repayment, shown in Fig. 10.4. The repayment process is when the system developer actually schedules and performs work to either remove the technical debt item or to reduce its impact on the health of the system. Repayment includes both restoring system performance affected by the technical compromise and mitigating long-term impacts on system health. Repayment can be performed against the principal, taxes, or interest of the technical debt item and can reduce impacts in any of the three system dimensions.

Since it is the system developer who is repaying the technical debt, we limit our discussion to the repayment of interest-bearing debt and repayment of the principal on fee-bearing debt. As identified in Sect. 10.1, plotting the balance versus the interest/fees of technical debt items enables the system developer to bin the items into four sections, each with a different repayment strategy. In this section, we will first cover these repayment strategies and then provide specifics for correcting technical debt.

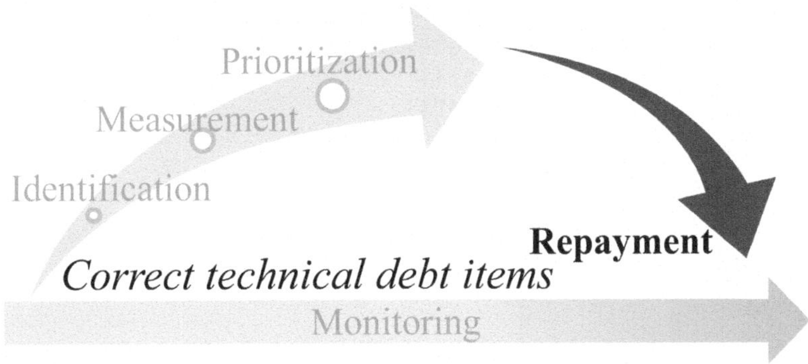

Fig. 10.4 Repayment process

10.2.1 Repayment Strategies

Applying the Eisenhower matrix (see Sect. 10.1) to technical debt items resulted in four categories of technical debt items, based on the urgency and importance of each item. Urgency is related to the level of interest or fees and is an indicator of how likely the technical debt item is to affect future system development or performance. Importance is related to the principal and taxes of the technical debt item and indicates how much effort is required to remove the technical debt item from the system. From the matrix, we develop a separate handling approach for each of the urgency/importance combinations: repay, mitigate, monitor, and track.

10.2.1.1 Repay

Technical debt items with high interest or fees and low principal are urgent, but not important. These items have a large impact on the development and use of the system. However, the actual correction required is small. Therefore, these items represent a large return on investment–a small amount of effort applied to correct the principal nets a large benefit throughout the system. The correct handling approach for these technical debt items is to repay the debt and correct the principal. By removing the technical compromise, benefits are rapidly seen throughout the system.

As an example, suppose that a software development team skipped peer reviews to release code on time and shipped the following code snippet, which was generated by Microsoft Copilot in response to the prompt “provide me an example of python code with a nested loop that results in poor runtime performance.”

```
#inefficient nested loop example
def check_common_elements(list1, list2):
  common_elements=[]
  for item1 in list1:
    for item2 in list2:
      #inefficiently checks all pairs
      if item1 == item2:
        if item1 not in common_elements:
          common_elements.append(item1)
  return common_elements
```

For those readers who are not software developers, this code is inefficient because it contains a nested loop—for every element in list 1, it checks every element in list 2. This means that the number of comparisons performed is the product of the size of the lists and therefore the time it takes to execute the code is related to that product. If this code was present in a released system, the execution time of the system increases significantly as the number of elements in the lists increases. This increase in the subroutine runtime could affect code that interfaces with this component and the efficiency of the entire program as experienced by the user. Fortunately, this code can be easily corrected and the correction (generated by Microsoft Copilot in response to the follow-on prompt "correct the inefficiencies") is shown in the next code snippet. In this code block, the first list is converted to a set, which enables rapid lookups of the entries and then a single for loop is run, iterating through list 2 and checking if the list items are in the set that corresponds to list 1.

```
#Optimized version
def check_common_elements_optimized(list1, list2):
  set1 = set(list1) #Convert a set for O(1) lookup
  common_elements = [item for item in list2 if item in
set1]
  return common_elements
```

Figure 10.5 shows the comparison in run time between the efficient (solid line) and inefficient algorithms (dashed line). The inefficient algorithm runtime increases exponentially, especially above 10,000 elements, while the efficient algorithm runtime stays relatively stable.

In today's world, it is not uncommon to have large data sets and to see this behavior. Algorithms that were originally coded for small data sets can exhibit poor performance when used on large data sets. This reuse is an example of architecture debt, specifically architectural lock-in (see Sect. 5.1). By reusing an existing code base and architecture with insufficient reviews of the code (the technical compromise), the system is inheriting performance issues which affect both the system developer (longer times to execute test cases and perform development) and the user (longer runtimes). The interest on the technical debt occurs when the developer tries to work around the inefficient code. Their debugging of additional modules may take extended time, and they may spend effort attempting to optimize the wrong part of

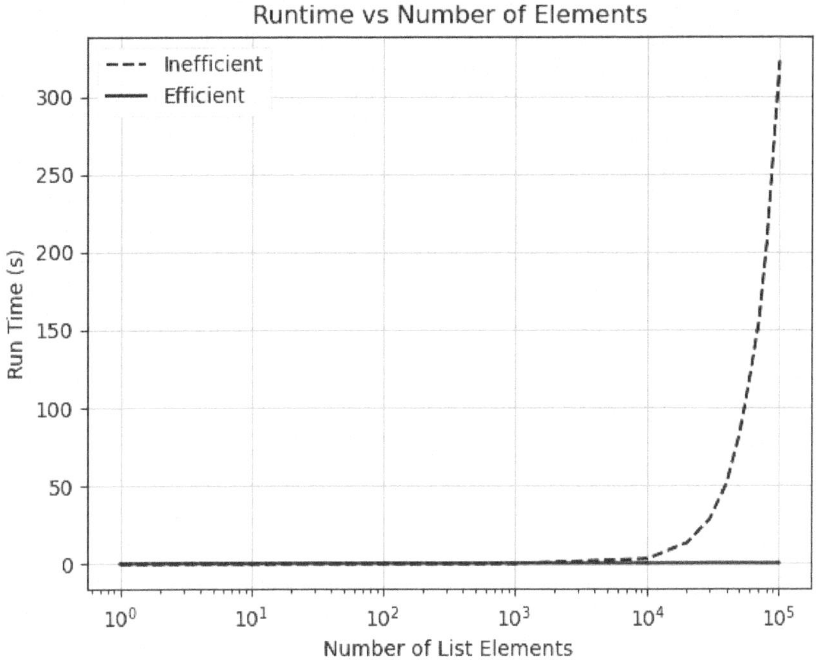

Fig. 10.5 Run time as a function of list size

the code. They may try to change the hardware specifications to run faster, adding material costs and affecting other system areas. A simple coding inefficiency, which was unintentionally introduced into the system, can cause significant interest and delays.

This example represents how some technical debt items may have low principal. In this case, the change is only a few lines of software code and the accompanying testing. However, the interest and fees can be large—as the list sizes increase the usability of the system drastically decreases due to the runtime impacts.

10.2.1.2 Mitigate

If the balance is large and the interest or fees are large, then the technical debt item is both important and urgent. Unlike the previous example, we can't just repay the debt because the balance size complicates the issue. Technical debt items with large balances will require significant rework and potentially introduce additional technical debt based on the rework.

Therefore, a more measured approach is required for these technical debt items. We call this approach "mitigate" because the system developer should develop a mitigation plan for these technical debt items which addresses both the interest/fee component and the balance component while working down the technical debt. This approach is akin to paying down a house mortgage—a measured payment each month over an extended period of time eventually reduces the debt. The initial payments pay more interest than principal. Over time, the share of principal paid with each payment increases.

So how can we develop a mitigation plan for a technical debt item? Similar to risk mitigation plans, a technical debt mitigation plan is a sequence of steps that the system developer takes to steadily decrease the interest, the principal, and/or the taxes. It is possible to mitigate fees through a patch, but by definition, that patch pays down part of the principal so we will not consider fee mitigation as a separate activity. We will consider three potential mitigation steps, one each for interest, taxes, and principal:

- Containing the spread of technical debt
- Implementing solutions that minimize taxes
- Mitigating principal.

The first thing to consider when mitigating interest is containing the spread of the technical debt throughout the system. If we let the technical debt "infect" new components that interface with the affected artifact, then those components will also experience the technical debt. Therefore, we must look to isolate the impacts of the affected portions of the artifact when connecting new components. In this way, the technical debt interest will not spread throughout the system. An example of containing the spread of technical debt is the "spaghetti code" and wrapper example provided in Sect. 7.2.2.

Taxes occur when the system developer is repaying the principal and has to rework a solution that was implemented due to the initial technical compromise. In essence, taxes are removing workarounds that were put in place to handle the existing technical debt in the system. To minimize the taxes requires minimizing workarounds by implementing wrappers and architectures that are flexible enough to support the solutions with and without the technical debt.

Finally, we have to mitigate the principal itself. These steps involve repairing the issues caused by the original technical compromise to restore the performance of the system. Simple examples include completing incomplete documentation and interface implementations. However, there are some situations where the principal may not be directly repayable—such as the switch from steel to aluminum in a car design. In these cases, the compromise cannot be "undone," however, work can be performed in other areas to regain the system scope or quality affected by the compromise. In the car case, aluminum may have a lower safety rating than steel due to structural properties, but other safety features, such as crumple zones, could be added to the design to minimize the impact on the safety rating. By adding additional safety features at the same time as the technical compromise is made, the total amount of principal is reduced.

Let's briefly discuss a few strategies on how to mitigate and repay technical debt. These strategies rely on the system developer to have identified the technical debt item and to have prioritized the technical debt register.

Isolate the Principal

Interest-bearing technical debt can be repaid by mitigating the continuing compounding of the interest or by paying down portions of the principal. Like a financial loan, paying down the interest is not as effective as paying down the principal. The existence of the principal will still cause additional interest to accrue, unless it can be fully isolated. Therefore, the first strategy for handling interest-bearing technical debt is to isolate the principal.

Isolating the principal involves working on the interfaces between the affected artifact and other system artifacts such that the aspects of the artifact affected by technical debt do not impact the interfaces. This goal can be accomplished by changing the affected artifact to repay the debt associated with the interfaces or by establishing containers in the connected components to shield them from further changes in the originally affected artifact. These containers limit the scope of changes associated with technical debt repayment and prevent the continued compounding of the technical debt.

Modular system design, with easily updatable and replaceable components is one way to isolate principal. As an example, consider a home computer. To keep the cost down, the computer manufacturer may build the computer with less random-access memory (RAM)—this is the technical compromise. The user will pay a technical fee of slower execution every time they run memory heavy applications. If the RAM is upgradeable, as is the case in many home computers, then the technical debt is isolated to the memory module. The user can upgrade the RAM if they are experiencing the technical fees, without impacting the rest of the system.

Continuous Repayment

With technical debt identified, it is up to the system developer to schedule time to gradually work down the technical debt. In a sprint-based development cycle, the system developer should reserve time each sprint for a few tasks related to paying down technical debt. This approach is similar to paying down a financial loan—you pay the same amount every month and eventually the loan is repaid. Similarly, the system developer can devote a few hours or days of effort every sprint to pay down the technical debt until it is corrected.

For this strategy to be successful, the developer must adjust their development schedules to account for technical debt repayment time. Without this accommodation, paying down the technical debt will result in unfinished work. The system stakeholders will not look favorably on the unfinished work which will create schedule pressure and cost pressure on the system developer, which will create more technical debt. Therefore, it is essential to get stakeholder buy-in on including technical debt repayment as part of the system development schedule.

Patch Releases and Hardware Servicing

There is often a scheduled cadence of releases for systems using an incremental or iterative delivery strategy. Updated versions may release once a quarter or once a year. A system experiencing technical fees can only provide corrections to their users through an updated release. Therefore, it may be necessary to release a patch (for software systems) or schedule a service visit (for hardware systems) to make the updates in between scheduled system releases.

Patch releases are simpler and easier to produce than a recall since the user can update the software themselves without needing to provide the developer with access to the system. Patches should be used to correct technical fees, when possible, instead of providing a hardware change. The developer should release patches when necessary, however, releasing patches too frequently may cause users to not trust the system or not to apply the patches.

It is less likely that the system user can perform the hardware updates required for the system, and therefore the developer may need to send out a service technician to make updates to address technical fees. While more complicated than software updates, hardware servicing can address the associated technical fees and also ensure that the rest of the system is properly maintained. Regular service calls and inspections can reduce the impact of fees by identifying maintenance needs prior to system downtime.

10.2.1.3 Monitor

If the balance on the technical debt item is large, but the interest and fees are small, then the technical debt item should be monitored. In this case, the technical compromise has resulted in a reduction in system performance that would be time-consuming to correct, but that compromise is not affecting the rest of the system development or is not likely to affect the rest of the development (low interest probability). Therefore, the system is in a relatively stable state and not deteriorating through the growth of interest or fees.

This type of technical debt must be regularly monitored by the system developer to ensure that interest does not start to accrue. Plans to repay the principal should be developed and attempts should be made to mitigate the principal as opportunities arise. These repayments are not as critical as those for technical debt items that also carry high fees or interest.

An example of technical debt that should be monitored is when the technical debt is contained to a specific component through isolation of the interfaces. As long as that component remains isolated then the likelihood that the high-principal technical debt causes high interest remains low, due to the low interest probability. Consider an efficient software module written in an old programming language, such as Fortran. The technical compromise is to leave the module in Fortran instead of updating it to a modern programming language supported by more developers. The principal would be the work required to update the module, and the interest is the work required to work around the module should its part of the system need to be modified. As long as the interfaces to the Fortran module remain isolated from the rest of the code and the hardware required to execute the module exists, then the interest probability on the technical debt remains low—there is little chance that the developer would need to work around this legacy component to deliver new functionality.

In this case, the principal is high—there are no longer a large number of Fortran programmers and the cost of redeveloping and reverifying the algorithm in another language could be large. The Fortran module must be monitored in case changes to the system environment occur, such as the retirement of the required hardware which would force the redevelopment of the module and repayment of the high principal. This example shows why it is important to monitor high-principal technical debt items. Even though they do not currently affect the system through large interest, they have the potential to significantly alter the progress of development should the principal need to be repaid.

10.2.1.4 Watch

The final category is for technical debt items with low principal and low interest and fees. These technical debt items fall under the watch repayment strategy, which means that they should be tracked but not acted upon. The low interest and fees indicate that there is little continuing impact on the system. Even though the principal is low, there is not a significant benefit to be had by repaying the principal. Therefore, these items should be tracked and, if time allows, repaid.

Watched items are different than monitored items. Due to the higher principal, monitored items require the development of mitigation plans. Watched items are simply tracked to ensure that there is no unexpected growth in principal, taxes, or fees. Technical debt that can be watched occurs when the technical debt is contained to a well-defined boundary within the system. For example, a router in a home network may be limited to a specific bandwidth and the owner may choose not to purchase a new router, making a technical compromise (router performance) to achieve a short-term cost savings. As new devices are added to the network, the overall bandwidth usage gets closer to the maximum supported by the router. Eventually, too many devices will be on the network and the router will need to be upgraded. Since the principal to replace the router is low (routers are not that expensive or hard to install), the interest is low (not a significant amount of extra work required to work around the old router), and the fees are low (degraded speed once too many devices are added), the technical debt item can be watched until the fees start to grow to the point where they affect the user (limited bandwidth to each device on the network).

10.3 Summary

In this chapter, we reviewed techniques for prioritizing known technical debt items and strategies for repaying the technical debt items. These two steps are critical to successful technical debt management as they reduce the total cost of the system technical debt, which reduces the likelihood of technical bankruptcy (see Chap. 12). Prioritization and repayment complete the actions in the management of a single technical debt item. However, the technical debt management cycle is a continuous process, and the system must be continuously monitored for new technical debt items. We'll discuss the monitoring process in the next chapter.

Reference

1. Wikipedia contributors. (2024, December 20). *Time Management.* Wikipedia, The Free Encyclopedia. Retrieved April 15, 2025, from https://en.wikipedia.org/wiki/Time_management#Eisenhower_method

Chapter 11
Monitoring Technical Debt

Abstract Monitoring technical debt is a continuous process within the technical debt management cycle. Monitoring includes observing the progress of existing technical debt items through the management cycle as well as identifying indicators of technical debt to assist in identification of new items. The technical debt review board is introduced and a method for integrating the technical debt review board with other systems engineering review boards is provided. The concept of the mission engineering and the mission systems engineer is introduced as a way to adapt to dynamic development environments. Multiple leading and lagging indicators of technical debt are discussed, including those related to schedule and performance.

Keywords Technical debt monitoring · Technical debt review board · Dynamic environments · Mission engineering · Technical debt indicators

The previous chapters in this part focused on managing a single technical debt item from identification to repayment. Since technical debt items can occur at any time in the system lifecycle, it is necessary to continuously monitor the system for technical debt to maintain system health. The monitoring process, shown in Fig. 11.1, occurs continuously and includes the following activities:

- Ensuring that technical debt management processes are followed
- Monitoring the progress of technical debt mitigations and repayments
- Tracking the evolution of the system environment to detect changes that affect system performance and,
- Engaging with stakeholders to manage the sources of pressure that cause technical debt.

The technical debt review board, discussed in Sect. 11.1, is responsible for the first two activities. The second two activities are more about reacting to the dynamic nature of system environments and stakeholders than managing system development. We discuss these two activities in Sect. 11.2. In Sect. 11.3 we discuss how to make technical debt part of the standard project lexicon to improve awareness and outcomes

H. Kleinwaks, *Managing Technical Debt in Systems Engineering*,
https://doi.org/10.1007/978-3-032-30360-8_11

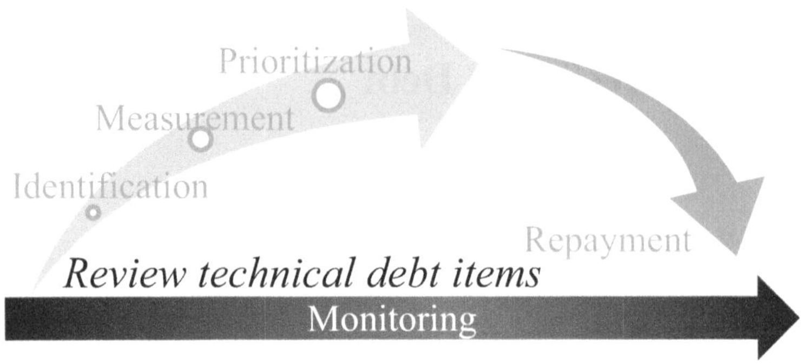

Fig. 11.1 Monitoring process

related to technical debt management. Finally, we conclude the chapter with a review of key indicators of technical debt within systems development.

11.1 Technical Debt Review Boards

Let's return to the Louis Gertsner quote we cited in Sect. 5.5: "People don't do what you expect but what you inspect" [1]. Inspection of the technical debt management processes occurs at the technical debt review board. The technical debt review board is chaired by the chief engineer and must include members that represent the system users (either actual users or proxies) such that potential technical fees can be identified. Other members of the board include the project manager, the mission systems engineer (see Sect. 11.2.2), and the technical leads for each major subsystem or component. The chief engineer chairs the board as it is necessary to ensure any system-wide performance impacts are well understood. However, the project manager must approve any mitigation efforts and their impact on project cost and schedule as part of the repayment process. Technical debt review boards should occur at least as often as risk management boards, however the frequency of technical debt review boards will be project-dependent and heavily influenced by the project duration and the current state of the system technical debt.

While each project should tailor their technical debt review board to meet their project specific goals, we recommend that the technical debt review board:

- Approves new technical debt items
- Monitors the progress towards mitigating technical debt items and,
- Ensures that the technical debt management processes are being followed.

New technical debt items can be proposed by any member of the development team as they are identified. The proposer of the technical debt item should craft the technical debt statement (Sect. 7.3.1) and generate all information required for the technical debt register (Sect. 7.2.4). The proposer then briefs the technical debt item to the review board, which decides to reject or approve the item. A proposed technical debt item may be rejected because it is insufficiently defined (such as not estimating the interest) or because it does not meet the definitions of technical debt (Sect. 3.5). If a technical debt item is rejected, then the proposer may bring it back to a future review board, assuming they have corrected the issues with it. Approving the technical debt item means that the item is included on the technical debt register and subject to the technical debt management process. It is important that every member of the team be encouraged to identify potential technical debt items, as they may be buried in a specific part of the system of which only a few team members have exquisite knowledge.

In addition to reviewing technical debt items raised by individual team members, each project should develop processes to automatically generate technical debt items raised by ERB/CCB decisions (Sect. 8.2.1.1) and by accepting risks (Sect. 7.2.3). These changes to the system have the potential for generating technical debt. Ideally, any generated technical debt is approved for entry into the register at the same time the ERB/CCB decision is made or the risk is accepted. Minimally, ERB/CCB and ROMB decisions must generate a technical debt item for review at the next technical debt review board such that the technical debt item can be included in the technical debt register and properly managed.

It is important to note that the technical debt review board does not approve the technical compromises that lead to technical debt. For intentional technical debt, approval of the compromises is the role of the ERB/CCB, ROMB, or another approval board within the organization. If the technical compromise is not acceptable, then this decision needs to be part of the ERB/CCB decision process: what engineering changes are required to come up with a different solution to generate acceptable (or not) technical debt? Similarly, if risk acceptance at a ROMB generates unacceptable levels of technical debt, then the risk should not be accepted. The technical debt review board ensures that the technical debt items are properly managed. For unintentional technical debt, the technical compromise has already been made prior to bringing the item to the review board. For these technical debt items, the technical debt review board will likely be the first place to bring attention to the potential consequences. The integration of the technical review board with other system review boards is shown in Fig. 11.2.

The technical debt review board is also responsible for monitoring the progress of technical debt item mitigation. At each board, updates to the interest, taxes, or fees of any technical debt item should be identified. Note that the principal is not on this list—it is measured at the time of technical debt item creation and is a static quantity. The progress of mitigation steps for each item undergoing repayment should be tracked and discussed and any required assistance identified. The technical debt review board decides if mitigation has progressed to a point sufficient to stop mitigating the technical debt item and to accept the potential consequences.

As discussed in Sect. 7.2.4, accepting technical debt items may create new risks. These risks are not the long-term consequences associated with the technical debt item. Instead, the acceptance of the technical compromise can introduce a new risk to the system—a new event that may or may not occur. As an example, consider

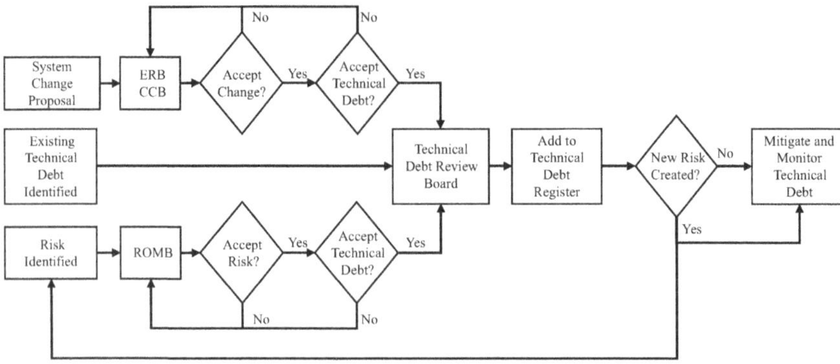

Fig. 11.2 Integration of the technical debt review board with other review boards

a technical compromise where a less performant component is accepted from a different supplier to save on cost (the short-term benefit). That supplier may not be as financially sound as the original supplier. By accepting the technical compromise, a new risk is created: the supplier may enter bankruptcy and therefore will not be able to fulfill its contract. The long-term consequences on the system associated with the technical debt relate to the performance of the component—it is known that a less performant component was selected. The risk is whether or not the supplier will be available to provide the required support—this event is uncertain.

The final job of the technical debt review board is to ensure that the technical debt management processes are being followed. Regular review boards enforce the processes by bringing forward newly identified technical debt items and ensuring they are measured. The review board prioritizes the new items along with the existing items and monitors the repayment progress. An example technical debt review board agenda is provided in Appendix C.

11.1.1 Tracking Technical Credit

Opportunities are the positive corollary to risks—an uncertain event that could have a positive outcome on the system. Similarly, technical credit is the positive corollary to technical debt. Technical credit occurs when the system developer makes a decision that may have a negative short-term consequence (such as increased cost) but a long-term benefit in the performance dimension. Just like opportunities are tracked at a risk and opportunity management board, technical credits should be tracked at the technical debt review board.

A system developer needs to carefully consider technical credits as they drive the system above its specified performance requirements. While over delivering can be beneficial to the users, it may result in unexpected cost or schedule overruns. These concerns are exactly why potential technical credit items need to be managed through the review board. Developers can bring their ideas to the review board, which can then carefully weigh the implications of enacting them and ask questions such as "Would implementing this solution delay the program past its milestones?" "What potential secondary or tertiary effects could this solution have and does it produce technical debt in another area?" These questions must be answered before a technical credit item is prioritized for action, and the technical debt review board is the proper place.

11.2 Handling the Dynamic Environment

Especially in today's digital world, systems are developed and deployed in environments that are volatile, uncertain, complex, and ambiguous (VUCA) [2]. Technological changes have significantly increased the pace of change. Moore's law is often cited as an exemplar of the rapid pace of technological advancement: the number of transistors on a computer chip doubles every two years [3]. This increase in computing power is substantial, and has enabled the development of increasingly powerful devices, to the point where the cell phones in our pockets have over 100,000 times more computing power than the systems used to send men to the moon on Apollo 11 [4]. What used to take computers the size of rooms can now be done in watches, enabling new ways of using technology and solutions. The rapid pace of change is incredible—ChatGPT was first released to the public in 2022, and as of this writing in 2025, it has become almost ubiquitous, changing how people write software code, read email, analyze reports, and perform other common tasks.

With this pace, it is nearly impossible for a system developer to predict the environment in which their system will be deployed—it will not be the same environment as when the requirements were written. This VUCA environment has driven system developers to more agile and adaptive system lifecycles and therefore there is an increased risk of technical debt. To minimize both interest and fees, the system developer must be able to handle the dynamic environment. In the following sections we'll address two approaches to handling VUCA environments: stakeholder management and mission engineering.

11.2.1 Stakeholder Management

Stakeholder management is a series of activities and processes by which the system developer works with their stakeholders to ensure that the stakeholder is properly informed about the system development progress. The system developer creates an open communication path to stakeholders such that they can address issues and ensure that stakeholder needs are met [5]. Establishing this open communication path allows the system developer to gain insight into emerging stakeholder concerns and needs and thus to get in front of potential technical debt.

A significant amount of technical debt starts with the pressure from system stakeholders who have specific business goals that need to be met. Budget limits and schedules are two of the major business goals. As we discussed in Chap. 4, the desire to satisfy these business goals can create pressure on the system developer and lead to the introduction of technical debt. The need to stay on budget and schedule is a fact of any project and should not be a surprise to the system developer and are items that the developer can track and manage. The key feature of stakeholder management as it relates to technical debt is staying aligned with the stakeholder and understanding when they will exert pressure due to observed system performance or new items of concern.

The open communication path with stakeholders can develop into a relationship such that stakeholders provide early indications of concerns with the system development progress, allowing the system developer to use project management techniques to adjust the progress instead of introducing technical debt. For example, if the stakeholder indicates that they are concerned about rising costs, the project manager can look for ways to reduce those costs without making technical compromises.

The other benefit of an open communication path with stakeholders is the ability to educate them on the potential technical debt and the consequences of introducing it into the system. Fully informing stakeholders allows them to understand the concerns of the development team and to make informed decisions, hopefully reducing the need to introduce technical debt into the system or enabling the introduced technical debt to be properly managed.

11.2.2 Mission Engineering and the Mission Systems Engineer

In the previous section, we examined how stakeholder management techniques can be used to respond to a dynamic environment and changing stakeholder needs during system development. However, the environment in which the system will be employed is also a VUCA environment. Rapid changes to the world state can change the application domain, and a system which cannot keep up will accrue domain debt. In this environment, the system may need to function as part of a greater system of systems (SoS). Decisions made only considering the single system may not appear to be a technical compromise or to have significant consequences, within the context of that system development. However, when viewed as part of the larger SoS, those same decisions may have significant impacts on the overall performance, meaning that technical debt has been introduced into the SoS.

The complexity of systems designs combined with the tendency of system developers to only focus on their specific components contributes to the failure of large integrated systems. As soon as system developers lose sight of the greater SoS and make decisions on their own, they introduce technical debt into the SoS that may not be recoverable. Let's look at an example.

Smart cities are cities that work to improve quality of life through the implementation and integration of real-time data feeds and digital technology. In order for the entire concept to work, multiple layers are required: a technology base layer which includes sensors and networks, an application layer to process and translate the raw data from the base layer, and a usage layer which represents the use of the applications by the city and its residents. Potential smart city applications and technologies include optimizing commuting, reducing crime, and improving public health [6]. However, for smart cities to provide these benefits, the entire SoS must work together to accomplish the mission.

If a technology developer is building a sensor to detect pedestrian traffic and makes a technical compromise that reduces the detection range, then the network of sensors will require additional sensors. Adding sensors exceeds the top-level budget, which then causes reductions in other parts of the smart city, and thereby impacts the health of the entire SoS. If the regulatory environment changes, then the locations where sensors are allowed to be placed may change, forcing different design decisions. If the sensors are aesthetically unpleasing, then they may not be able to be placed in highly public and trafficked areas. These considerations are beyond the purview and control of the individual sensor developer but are no less important to the overall success of the system.

This example begs the question: how does a system developer stay aware of the changing environment and the evolving SoS? The developer must be aware of the role their system plays in the greater SoS and the relevant figures of merit and performance metrics. This is a complicated task and requires a specialized engineering framework and engineer within the development team to represent the interests of the SoS. This framework is defined within the discipline of Mission Engineering.

The United States Department of Defense defines Mission Engineering as "an interdisciplinary process encompassing the entire technical effort to analyze, design, and integrate current and emerging operational needs and capabilities to achieve desired mission outcomes" [7]. Mission engineering focuses on end-to-end analysis of the SoS to inform design decisions across technical, personnel, support equipment, and policy areas. The goal of mission engineering is to ensure that the "right things" are delivered to achieve mission outcomes. Mission engineering provides the inputs to the systems developer to help validate the system by specifying how the system will be used within the greater SoS construct and communicating the critical aspects of system performance. Mission engineering activities should be initially completed as part of the needs analysis stage and continuously refined throughout system development. The results of the mission engineering process are communicated to the system developer and kept up to date based on changing mission needs.

The system developer uses the results of mission engineering to establish validation criteria for the system. To understand this process, we first need to define a mission. The mission engineering guide defines missions as "tasks and action undertaken to achieve a specific objective" [7]. In a broader context, a mission can be viewed as the specification of a use case for the SoS. The use case defines the sequence of tasks required for the SoS to function in a specific way. Note that we are referring to the use case at the SoS level and not the system level. This is a critical distinction, as the goal of mission engineering is to enable each system to do its part to enable the overall mission success and not on the exact functionality of the individual system. This focus on SoS behavior and mission success enables the evaluation of each system within the greater SoS context and different conditions. The ability of the system to perform as part of the SoS determines its suitability for a mission.

Figure 11.3 shows an example output of an integrated mission analysis, showing how different potential solutions are better at different times throughout the scenario.

This analysis can be used to evaluate the effects of technical tradeoffs within a single system on overall mission performance. A decision in the design of the system may move the performance from the baseline solution to alternative 2. Alternative 2 performs better at some points in the mission, but not at all points, including a trailing off once the variable goes above 18, at which point alternative 1 performs better. The variable is determined based on mission context, which impacts the choice of the alternative, and therefore the overall mission context is critical to evaluating the effects of technical tradeoffs.

Understanding both the system under development and the mission context is complicated. In these types of projects, it is beneficial to assign a mission systems engineer the responsibility for ensuring that the delivered system will perform its required role within the greater SoS. This means that the mission systems engineer must be aware of both the mission and the system and the interconnections between the two. The mission systems engineer must be able to model system performance within the mission context to ensure that the system can be validated within the intended application domain.

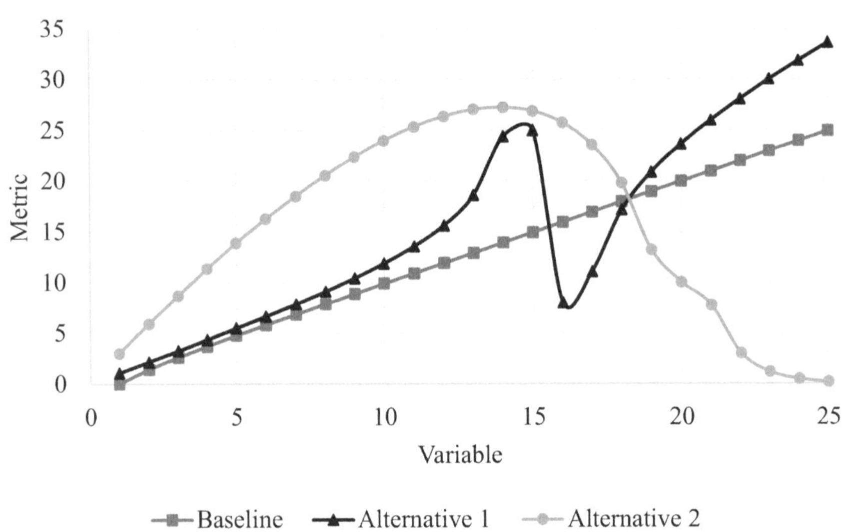

Fig. 11.3 Notional mission engineering analysis outputs

Therefore, the mission systems engineer is responsible for evaluating technical compromises and technical debt for its impact on the mission. Technical compromises made at the system level may seem to have insignificant impact until they are viewed within the greater mission context. It is the mission systems engineer's job to bring the mission context to the technical debt review board and to assess technical compromises, focusing on limiting incurred domain debt. Let's look at an example of a mission definition outside of the context of the DoD to show how the mission systems engineer is responsible for limiting technical debt.

Consider a company that makes and installs security systems. Each security system is a SoS consisting of cameras, motion sensors, alarms, and other components. Each of these components is its own system: they collect data, distribute that data to a central processor, and are subject to constraints such as power and lighting conditions. One mission of the security system is to detect an intruder, notify the police, and provide the police with detailed video to enable identification of the intruders, which requires the various systems to work together as a SoS. Each security installation is unique to the building in which it is installed, giving each installation a different mission context. Figure 11.4 shows a notional security system setup for a bank.

During development of this security system, the stakeholder decides that the project is too expensive and they need to cut costs, opting for a cheaper camera system. Initially, this trade seems acceptable. Describing the mission and the use of the system within the mission, however, reveals the technical debt. Part of the mission is to enable identification of the user by running facial recognition on the video. While the cheaper camera can produce the proper resolution for facial recognition software, it can only do so from a distance of 10 ft or less. In the installation shown in Figure 11.4, the camera is mounted 20 ft from the doorway and 15 ft from the bank counter, the primary locations which the intruder would access. Therefore, although the camera met the requirement to produce high resolution video, its range limitation would not allow the system to meet its mission an intruder may walk in and never be close enough to the camera for facial recognition to be attainable. Another camera location, more cameras, or a better camera is required. The mission

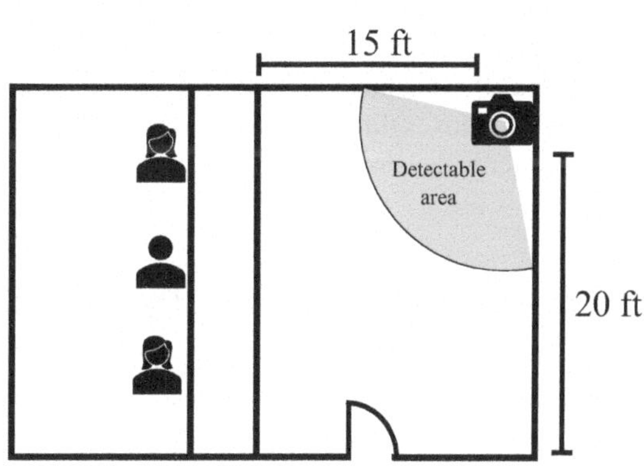

Fig. 11.4 Notional bank security system setup

systems engineer needs to understand how the systems work together to ensure that the mission can be completed and that technical debt accepted by one system does not adversely affect the success of the mission.

11.3 Updating the Lexicon

For technical debt management to be effective, we need to ensure that the potential for technical debt is considered by all members of the development team during their decision-making processes. Doing so requires that technical debt terminology become part of the standard lexicon used by the team.

We know that the technical debt ontology is not commonly used in the field [8]. By introducing the processes outlined in the previous chapters and the technical debt review board, this can change. I introduced the concept of technical debt to one of my spiral development projects over the last two years. I provided training for each team member on the definition of technical debt and on identification methods. I then identified several existing technical debt items. By bringing up technical debt repeatedly in various reviews and forums, I was able to make technical debt part of the everyday lexicon. Now, every ERB decision considers technical debt and we discuss the potential for technical debt when pressure from our stakeholders forces technical compromises.

Making technical debt part of the standard project lexicon increased our awareness of it. We actively discuss the consequences of our decisions and have even appointed a technical debt manager. It is surprising how little effort it takes to bring the concepts of technical debt to the forefront of decision making. By simply using the terminology, the awareness of the potential for technical debt increases. Increasing the awareness of all team members reduces the unintentional technical debt in the system and the consequences of intentional technical debt, moving from reckless to prudent debt.

11.4 Indicators of Technical Debt

We conclude this chapter by looking at specific indicators of technical debt that occur as the system moves through its lifecycle. These indicators represent common issues seen in systems development.

11.4.1 Sprint Carryover and Delays in Task Completion

In sprint-based lifecycles, the work assigned in each sprint is supposed to be of small enough scale to be completed within that sprint. At times, the work does not complete and carries over to the following sprint —the work is late. The carryover may be due to multiple reasons, including late dependencies or unanticipated complexity. Delays in task completion in other system lifecycles are also potential indicators of technical debt. Each delayed task reduces the value of the system through reductions in profit or schedule. If these delays continue, the stakeholders will respond by pressuring the developer to release on time or budget, introducing technical debt.

The system developer should closely follow the progress of tasks, especially in adaptive lifecycles, and should institute corrective techniques early to prevent continued delays. Otherwise, the system developer will be forced to take shortcuts later in the program, such as minimizing testing, to recover lost time or money. These shortcuts lead to technical debt. Sprint carryover and delays in task completion are both leading and lagging indicators of technical debt. They are leading indicators when the task delays may force developers to take shortcuts to complete them and they are lagging indicators when the task delays are due to increased complexity from previous tasks.

11.4.2 Minimizing Tests in Favor of Analysis or Inspection

During the early stages of the system lifecycle, it is common for the developer to produce a RVTM, which maps the requirements to their verification events and defines the verification methods to be used. This matrix, typically first reviewed at the SRR, provides early indications of requirements debt and test debt. An example of an RVTM for an electric vehicle charging station is shown in Fig. 11.5. In this RVTM, the requirement number and its source document are specified along with the text of the requirement. Then the verification method, the event when verification will occur, and the verification evidence are listed. Finally, a status column tracks the state of the requirement verification.

The verification method for a requirement is typically one of the following [9]:

- Analysis: models, simulations, and analytical computations are used to determine if the system will meet the requirements. Frequently used for requirements that

Requirment Number	Document	Text	Verification Method	Verification Event	Verification Evidence	Status
1.1	System Requirements Specficiation	The system shall charge a Hyundai IONIQ 5 to 80% charge within 20 minutes	Test	Charging Test	Charging test report	Not Started
1.2	System Requirements Specficiation	The system shall concurrently charge 3 vehicles.	Inspection	Design Reviews	Design Drawings	Complete
1.3	System Requirements Specification	The system shall display the state of charge to the user while charging.	Demonstration	Display Demonstration	Demonstration Results	Not Started

Fig. 11.5 Snippet of an example RVTM

cannot be easily tested or where the testing is too expensive. Similarity (comparing to a similar system) is a type of analysis that is also sometimes used for verification.
- Demonstration: The system is shown to perform in its intended environment, but without taking measurements and data. The system is stimulated and its reactions are observed and compared to the expected behavior.
- Inspection: Visual review of the system or its documentation, typically used for simple requirements that are best verified by observation, such as color or number of elements, or by simple measurements, such as the mass of a system component.
- Test: A measured execution of a system function in a controlled environment, where data is collected and assessed. Test often requires supporting equipment and instrumentation.

Of these methods, test is the costliest and the most time consuming, often requiring specialized test equipment and data analysis. System developers may want to reduce the number of requirements verified by test such that the schedule can be compressed and the budget reduced. However, an over-reliance on inspection and analysis can lead to test debt—the system is not properly tested and therefore does not work in its intended environment. One of my (least) favorite examples of the overuse of inspection is when the system developer states that they will verify compliance of a software component to an ICD by inspection. The fact that the ICD is delivered can be verified by inspection, however, the interface must be tested (with the components on both sides of the interface) to prove that the system complies with the ICD.

The RVTM also provides insight into the system developer's understanding of the requirements. The RVTM may not use proper verification methods or verification events because the system developer does not understand the requirements or because the requirements are not well specified. Therefore, the RVTM can be a leading indicator of requirements debt when verification methods and events do not align with stakeholder interpretation of the requirement.

11.4.3 Excessive Parallelization

Some tasks within the development of a system can be easily parallelized–they do not depend on each other and they use different resources. Other tasks cannot be parallelized due to dependencies. A part cannot be manufactured until the design is complete and a system cannot be fully tested until all of its component parts are delivered and integrated. To shorten schedules, project managers may start a task before its predecessor is fully completed. This approach is logical if one considers the S-curve shape of project earned value [10]. In this view, the earning last few percent of a task's value takes significantly longer than achieving the first 90% of the value. Figure 11.6 shows a curve where the last 10% of the value takes 35% of the time. Therefore, if those last few percentage points of task performance are not necessary to the following task, why not start the task sooner? Research has shown that value delivery can be modelled with an S-curve and that starting tasks earlier did allow for earlier overall completion [11].

Scheduling overlapping tasks may seem advantageous in shortening schedules, but it also runs the risk of increasing technical debt. If that last 5% of the work on the predecessor task changes something in the interface used by successor tasks, then the successor tasks will need to be reworked. If something in that last 5% is needed by the successor task, then the successor task may make erroneous assumptions, inserting additional technical debt into the system.

Independent tasks may be executed in parallel with no relationship between them, so they are not a concern here. For tasks with dependencies on other tasks, there are typically four types of relationships [12]:

- Finish-to-Start: the predecessor task must finish before the successor task starts
- Start-to-Start: the predecessor task must start before the successor task starts
- Finish-to-Finish: The predecessor task must finish before the successor task finishes
- Start-to-Finish: The successor task must start before the predecessor task finishes.

Finish-to-Start tasks are not parallelized. The other three relationships do indicate some level of parallelization. If we are creating the relationship in the schedule then personnel, components, documentation, or some other resource must pass from one task to the next. If this resource is not fully developed at the time the successor task starts, then we are likely to incur technical debt. Therefore, the use of Start-to-Start, Finish-to-Finish, and/or Start-to-Finish relationships in a schedule is a leading indicator of potential technical debt.

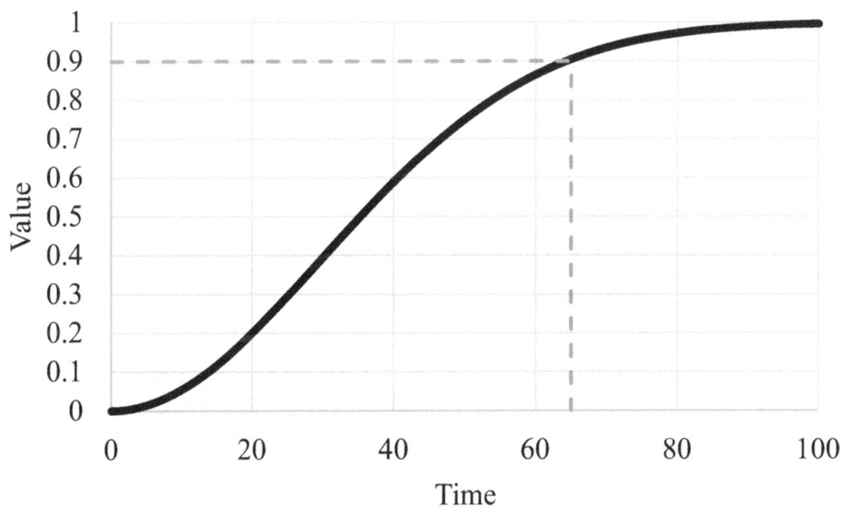

Fig. 11.6 S-curve of value creation

11.4.4 Insufficient Modeling Fidelity

Modeling and simulation are used to predict the performance of a system. Modern digital tools allow system developers to rapidly adjust parameters and revaluate performance prior to finalizing the design and building hardware. When building models and simulations, the level of fidelity is important. Modeling and simulation fidelity is captured in the modeling pyramid as shown in Fig. 11.7. The lower levels of the pyramid are higher fidelity models, which tend to take longer to develop and execute. At the lowest level are the physics and engineering models which model the physical laws impacting the system and generally apply to the system components. The next level models system-wide effects which look at the entirety of system performance, without accounting for its application within the architecture or the environment. The third level models the overall architecture and evaluates the effectiveness of the system within its intended environment and its interactions with other systems. Finally, the top of the pyramid models mission level use cases across multiple campaigns.

The approach in Fig. 11.7 is valid for any system. A robotaxi maker needs to have models of its power train (the physics/engineering level), the performance of the robotaxi (system level), the interaction of the robotaxis with other vehicles, weather, and the road environment (architecture level), and plans for maintaining and deploying the fleet of robotaxis (the mission level). Similarly, a missile manufacturer needs physics models of the missile propulsion system, system models of the missile against various targets, architecture models of the missile as part of an aircraft fighter engagement, and mission models of the missile performance in an integrated air attack and defense campaign.

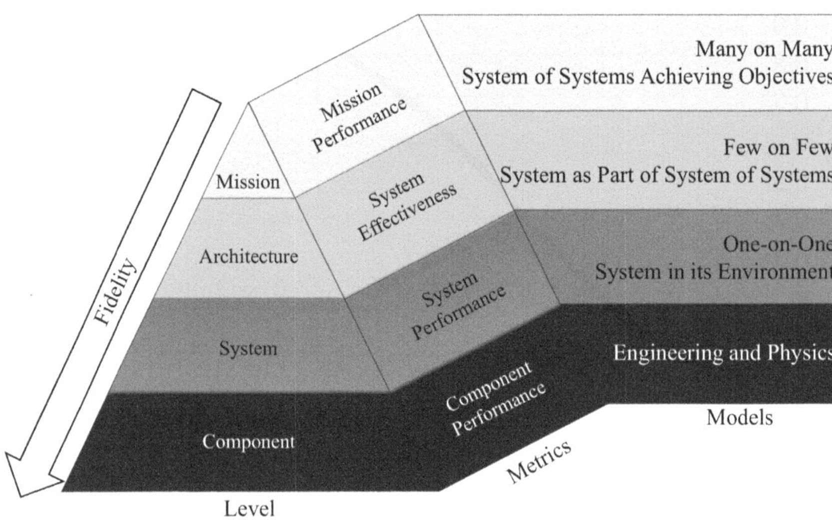

Fig. 11.7 Modeling and simulation pyramid, adapted from [13]

The benefit of the modeling and simulation pyramid is that results from lower levels can be used to abstract model components at higher levels. If the physics, engineering, and systems models show that the robotaxi can operate for five hours on a single charge, then that operating time can be used in the architecture and campaign level modeling. However, if the specified five hours are never verified through modeling and simulation of higher fidelity, then the system has incurred modeling and simulation debt. The system developer must make sure that modelled numbers are verified and that modeling and simulation is done to the level of fidelity required for the specified application.

Model fidelity is a leading indicator of technical debt. If the models are shown early to be insufficient to assess performance of the system, then there is a high likelihood that the system will not perform as expected. The technical compromise on the model, which includes the use of unfounded assumptions, will result in poor performance predictions, causing technical fees.

11.4.5 Incomplete Interface Specifications

An interface occurs at the boundary of any two system components [9], whether the components are hardware or software components, and at the boundary of the system and its environment. When components are developed independently, there must be a specification to control the interface between them and to ensure that the integrated set of components will perform as intended. These interface control documents (ICD) are a common source of documentation debt within the system.

To accelerate development schedules, system developers may begin work on a component before its interfaces are completely specified. While enough of the ICD may be complete to start the work, technical debt is accrued because there is the potential for future ICD changes to force rework on the component. One side of the interface may develop part of their system that is incompatible with the other side of the interface, but this incompatibility is unknown at the time since the ICD was incomplete.

Another source of documentation debt from ICDs and other technical standards occurs when the documentation leaves room for the developer to misinterpret the meaning in its implementation. Typical sources of confusion include optional parameters and "or" statements in the specification. As an example, consider the digital video broadcast second generation standard (DVB-S2) used to control the format of radiofrequency transmissions between satellites and their ground stations. The specification states "Data services may be transported in Transport Stream format according to EN 301 192 (e.g. using Multi-protocol Encapsulation), or Generic Stream format" [14]. This statement in the standard can be interpreted to mean that a system only needs to implement the transport stream format or the generic stream format and not both.

If the two ends (satellite and ground station) of the radiofrequency link select different stream formats to implement, then the systems will both be compliant to the standard but will still be incompatible! This occurrence is an example of documentation debt due to an interface specification left open to interpretation. We like to think that ICDs and standards are more proscriptive than this example, but unfortunately, that is not the case–each requirement in the ICD tends to be open to interpretation. This interpretation is why interfaces that implement ICDs must be tested to ensure compliance instead of singly evaluating one side of the interface.

Incomplete, incorrect, or invalid interface specifications and ICDs are a lagging indicator of technical debt. By the time it is determined that the specification is insufficient, the technical debt has already entered the system. The use of "to-be-determined" (TBD) within an ICD or specification is a leading indicator of technical debt. The system developer will work towards that specification, but without final values they may design a system which is incompatible with the interface once it is completed. Starting development work before the completion of the ICD is also a leading indicator of technical debt–the schedule compromise to begin work can result in erroneous assumptions followed by rework once the ICD is completed.

11.4.6 Changes to the System Environment, Including Changing Stakeholder Needs

Agile and adaptive development strategies were created to deal with dynamic environments. By frequently updating the stakeholders on system development and receiving real-time feedback, developers can adapt the system design concurrently with changes in stakeholder needs. However, this adaptation tends to focus on short-term goals and can introduce technical debt into the system. Therefore, domain debt can be created and the final system may not perform as expected in the operational environment.

Stakeholders with changing needs can also be an indicator of technical debt in the system. A stakeholder with frequently changing needs will put different pressures on the system developer which will result in multiple technical compromises. These changes show up as updates to requirements, contract modifications to add scope, and inconsistent feedback at system reviews. A known dynamic environment is a leading indicator of technical debt–as the domain changes, the system will be hard-pressed to keep up with it.

11.5 Summary

This chapter finishes our discussion of the technical debt management cycle. Monitoring technical debt is a continuous process essential to proper technical debt management. We provided mechanisms on how to monitor technical debt, including technical debt review boards, responding to dynamic system development environments, and updating project lexicon to increase awareness of technical debt. Finally, we provided several indicators of technical debt that can be monitored to help find unintentional technical debt that is already in the system or that may soon be inserted into the system. Specifically, we identified the following leading and lagging indicators of technical debt:

- Leading indicators: sprint carryover and task delays, insufficiencies in an RVTM, excessive task parallelization, model fidelity, and known dynamic environments
- Lagging indicators: sprint carryover and task delays, and incomplete or invalid ICDs.

The goal of managing technical debt is to keep the system healthy and reduce the risk of technical bankruptcy. In the next chapter, we will dive deeper into technical bankruptcy to establish how it occurs and how to avoid it, and we will provide an example of a large system that experienced technical bankruptcy, resulting in project cancellation.

References

1. Kimberley, O. (2025, March 10). *It's Time For Marketers To Redefine The RFP As The 'Request For People'*. Forbes. Retrieved April 7, 2025, from https://www.forbes.com/sites/quad/2025/03/10/its-time-for-marketers-to-redefine-the-rfp-as-the-request-for-people/
2. Kleinwaks, H., Batchelor, A., & Bradley, T. H. (2023). Technical Debt in Systems Engineering - A Systematic Literature Review. *Systems Engineering, 26*(5), 675–687. https://doi.org/10.1002/sys.21681
3. Wikipedia contributors. (2025, April 3). *Moore's law*. Wikipedia, The Free Encyclopedia. Retrieved April 23, 2025, from https://en.wikipedia.org/w/index.php?title=Moore%27s_law&oldid=1283756153
4. Kendall, G. (2019, July 9). *Apollo 11 anniversary: Could an iPhone fly me to the moon?* The Independent. Retrieved April 23, 2025, from https://www.independent.co.uk/news/science/apollo-11-moon-landing-mobile-phones-smartphone-iphone-a8988351.html
5. Project Management Institute, Inc. (2017). *A Guide to the Project Management Body of Knowledge (PMBOK Guide)* (6th ed.). Project Management Institute, Inc.
6. Woetzel, L., Remes, J., Boland, B., Lv, K., Sinha, S., Strube, G.,... & von der Tann, V. (2018). *Smart cities: Digital solutions for a more livable future.* McKinsey Global Institute. Retrieved from https://www.mckinsey.com/capabilities/operations/our-insights/smart-cities-digital-solutions-for-a-more-livable-future
7. United States Department of Defense, Office of the Under Secretary of Defense for Research and Engineering. (2023). *Department of Defense Mission Engineering Guide.*
8. Kleinwaks, H., Batchelor, A., & Bradley, T. H. (2023). An Empirical Survey on the Prevalence of Technical Debt in Systems Engineering. *INCOSE International Symposium, 33*(1), 1640–1658. https://doi.org/10.1002/iis2.13104
9. Walden, D. D., Roedler, G. J., Forsberg, K. J., Hamelin, R. D., & Shortell, T. M. (Eds.). (2015). *Systems Engineering Handbook* (4th ed.). John Wiley and Sons.
10. Warburton, R. D. (2011). A time-dependent earned value model for software projects. International *Journal of Project Management, 29*(8), 1082–1090. https://doi.org/10.1016/j.ijproman.2011.02.008

11. Kleinwaks, H., Batchelor, A., & Bradley, T. H. (2023). Predicting the Dynamics of Earned Value Creation in the Presence of Technical Debt. *IEEE Access, 11*, 125381–125395. https://doi.org/10.1109/ACCESS.2023.3331112
12. Jones, E. F. (2009). Scheduling 101—the basic of best practices. *PMI® Global Congress 2009*. Project Management Institute. Retrieved from https://www.pmi.org/learning/library/schedule-101-basic-best-practices-6701
13. Air Force Research Laboratory. (n.d.). *AFRL/RY - COMPASE - Modeling and Simulation*. Air Force Research Laboratory. Retrieved April 26, 2025, from https://www.afrl.af.mil/About-Us/Fact-Sheets/Fact-Sheet-Display/Article/2331811/afrlry-compase-modeling-and-simulation/
14. European Telecommunications Standards Institute. (2014). *Digital Video Broadcasting (DVB), Second generation framing structure, channel coding and modulation systems for broadcasting, interactive services, news gathering and other broadband satellite applications (DVB-S2)*. (ETSI EN 302 307 V1.2.1 (2009-04)).

Chapter 12
Technical Bankruptcy

Abstract The concept of technical bankruptcy is introduced and methods provided to quantify it and reduce the risk of it occurring. A system reaches technical bankruptcy when it cannot proceed with its lifecycle without receiving additional resources (increase budget or delayed schedule) to correct some or all of the existing technical debt items. This condition occurs due to the accumulation of unmanaged technical debt within the system. Three conditions are presented which create technical bankruptcy: the cost to continue working around the technical debt is higher than the technical debt balance, the lesser of the workaround cost and the technical debt balance is higher than the available margin, and the total fees exceed the performance threshold acceptable to the user. These conditions are used to derive methods to reduce the risk of technical bankruptcy. The JMS program is presented as an example of a system that experienced technical bankruptcy.

Keywords Technical bankruptcy · Estimating technical bankruptcy · Mitigating technical bankruptcy

Technical bankruptcy is the state where the system cannot proceed with its lifecycle until some, or all, of the technical debt is repaid. Technical bankruptcy may occur when the system development exceeds the available budget or schedule due to increased development required to account for technical debt [1]. Technical blockers may also cause technical bankruptcy when workarounds are no longer feasible and rework is required [2].

Technical bankruptcy sits on the left side of the technical debt context map, as shown in Fig. 12.1. The health of the system contributes to technical bankruptcy by impacting the satisfaction of business goals. If those goals cannot be met within the project budget and schedule, then the system may be technically bankrupt. Within this chapter, we will look deeper into technical bankruptcy to establish how a system reaches this state and how the system developer can recover.

H. Kleinwaks, *Managing Technical Debt in Systems Engineering*,
https://doi.org/10.1007/978-3-032-30360-8_12

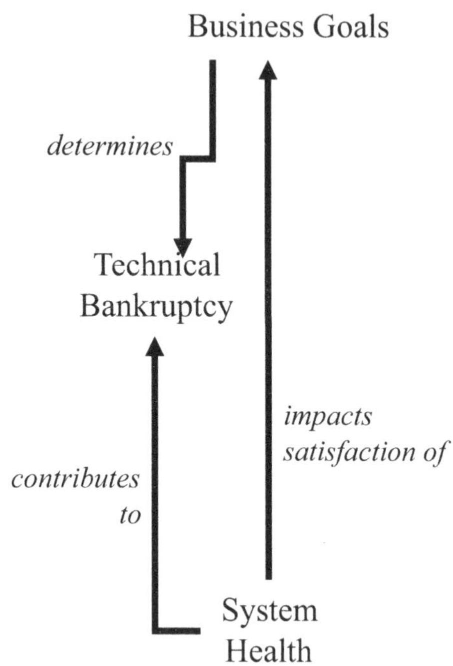

Fig. 12.1 Technical bankruptcy in the technical debt context map

12.1 Moving from Technical Debt to Technical Bankruptcy

Technical bankruptcy occurs when the system accumulates too much technical debt. In financial bankruptcy, the borrower cannot stay in business until they pay back their creditors. The same concept applies in technical bankruptcy: the system cannot continue in its lifecycle until some or all technical debt is repaid. How does technical debt accumulate to the point where the system lifecycle can no longer proceed?

As we have seen throughout this book, technical debt occurs when a compromise that is made in the performance dimension to achieve a short-term benefit negatively affects the long-term health of the system. The negative impacts can be in any of the system dimensions, and therefore can accumulate across all dimensions. Accumulating technical debt reduces the value of the system by lowering the values in each dimension. Eventually, the value is reduced to the point where the stakeholder demands that changes are made to restore the value. This restoration comes by spending time or money to restore performance by repaying the technical debt.

When left unmanaged, the accumulation of technical debt creates technical bankruptcy. When this occurs, the system cannot proceed with its lifecycle. Symptoms of technical bankruptcy include:

- Exceeding project budget and requiring additional funds;
- Exceeding project schedule requiring a rebaselining of the schedule and extra time to deliver;

- Failing gate reviews due to system underperformance;
- Technical fees that prevent the user's acceptance of the system; and,
- Project termination.

It is important to note that reducing the profit or delivering late does not mean that the system has entered technical bankruptcy. Bankruptcy occurs when the values are exceeded to the point that the development cannot continue. For example, a system developer may lose profit on a project but still deliver—that is not bankruptcy. Instead, bankruptcy occurs when the system developer can no longer afford to continue developing the project. Therefore, each project has a unique threshold for technical bankruptcy. In the next section, we'll talk about how to quantify technical bankruptcy.

12.2 Quantifying Technical Bankruptcy

Given that the thresholds for technical bankruptcy are unique for each project, we cannot provide a single value of accumulated technical debt that denotes technical bankruptcy. Instead, we will discuss how to use the technical debt measurement concepts introduced in Chap. 9 to establish the overall technical debt load. By quantifying the technical debt load, the system developer can work with stakeholders to understand the risk of technical bankruptcy within the system.

One example of the quantification of technical bankruptcy is the Nunn-McCurdy breaches within the United States Department of Defense (DoD) . DoD programs experience a Nunn-McCurdy breach if the budget for the program, measured by the program acquisition unit cost or average procurement unit cost, exceeds the current baseline estimate by 25% or exceeds the original baseline estimate by 50%. At this point, the program must be terminated unless the Secretary of Defense intervenes with Congress [3]. Nunn-McCurdy breaches have been caused by scope changes, misunderstandings of requirements, and infeasible assumptions about development timelines and budgets [4]. When a Nunn-McCurdy breach occurs due to increasing technical debt, it is an example of technical bankruptcy. Table 12.1 provides an overview of several Nunn-McCurdy breaches and the potential role of technical debt in each one.

Table 12.1 Role of technical debt in 2011 Nunn-McCurdy breaches [4]

Program	Potential role of technical debt
Army warfighter information network-tactical communications system	Infrastructure debt: COTS radio technology incorrectly assumed to be available forcing unique development activities
Space-Based Infrared System High	Design debt: underestimated amount of software that would be able to be reused instead of redesigning the entire system
Excalibur	Domain debt: reduction in desired quantity of munitions to be purchased, causing the per unit cost to exceed thresholds, due to a misunderstanding of the intended domain and use

Technical bankruptcy stops the system development from continuing—what does that mean for the cost to correct the technical debt compared to the cost to work around the technical debt? If the technical debt is easy to work around, then the system can continue in its lifecycle even if the cost to correct the technical debt is high. However, if the cost to work around the debt is significant, then the system developer will be faced with a choice: continue to experience the large cost of workarounds or repay the balance to correct the technical debt. It is this situation that can prevent the system development from continuing, and therefore we have our first condition for technical bankruptcy: the continued cost of working around the technical debt (the interest) must be higher than the cost to repay the technical debt (the principal and taxes).

Technical Bankruptcy Condition 1: $\overline{C_I} > B_t$

In condition 1, it is important to note that the costs are expected values and expressed in current day terms. This condition is satisfied when the expected value of the cost to continue to work around the technical debt is greater than the expected value of the cost to repay the technical debt. As components are added to the system, interest is paid to integrate them with existing components which bear technical debt, often in the form of longer development timelines. However, once the components have been added, that part of the interest is paid, leaving behind a system which still has technical debt and in which the taxes (the cost to remove the technical debt) have increased. If the new component used a workaround that created additional technical debt, then there is the potential for continued interest growth.

To determine if condition 1 is satisfied, we must examine both the technical debt balance and the workaround cost. The balance, which consists of the principal and taxes, is calculated using Eq. (9.8), which is repeated here:

$$B_t = \sum_{j=0}^{n} \left(P_j + \sum_{i=0}^{m} T_{t,j,i} \right) \tag{9.8}$$

The workaround cost $\overline{C_I}$, shown in Eq. (12.1), is the expected cost of the interest, which is the interest terms from the total cost in Eq. (9.9)

$$\overline{C_I} = \sum_{j=0}^{n} \left(\sum_{i=0}^{n} a_{t,j,i} * r_{t,j,i} \right) \tag{12.1}$$

If the workaround cost exceeds the repayment cost, then the system developer should repay the technical debt first. Repaying technical debt does not move the system forward in its lifecycle and therefore, if the total cost of working around the system technical debt is higher than repaying the system technical debt, then the system has reached technical bankruptcy. When some of the system technical debt balance is repaid, it also reduces the cost to work around the debt since it removes the principal, taxes, and interest. Therefore, repaying technical debt items can rapidly move the system out of bankruptcy.

The next condition that can cause technical bankruptcy is when either the workaround cost or the repayment cost exceeds the project margins.

Technical Bankruptcy Condition 2: min$\left(\overline{C_I}, B_t\right) > Margin$

Condition 2 occurs when the system developer must choose to either work around the technical debt or repay the technical debt and either option exceeds the available resources, also known as the margin. In this condition, the developer does not have the budget or schedule (the resources) to implement the workaround or repay the balance and therefore the project will stop progressing through its lifecycle unless additional resources are added. Establishing condition 2 requires relating $\overline{C_I}$ and B_t to the available resources and knowing the limit of those resources.

Recall from Chap. 9 that principal, taxes, and interest are all measured as duration-based values: the time to repay the technical debt (principal and taxes) and the increased time to perform work (interest). To evaluate condition 2, these values are compared to the margin and therefore the margin should also be identified in duration units: the amount of additional time that the stakeholders are willing to spend developing the system. Alternatively, the durations can be converted to monetary values and the margin can be expressed as a budget.

Finally, we must look at fee-bearing debt and the total fees, F_t. Recall that the total fees are calculated using Eq. (9.10):

$$F_t = \sum_{i=0}^{n} g_i * f_i * t \tag{9.10}$$

The total fees can be thought of as the usage cost—the total cost to the user compared to a system that did not have technical debt. Technical fees are paid by the user and do not extend the project timelines. However, excessive fees can result in the system failing validation and being rejected by the user. Such failures either result in significant rework of the system or in project cancellation. Thus, technical bankruptcy condition 3 occurs whenever the usage cost is greater than the user's threshold limit.

Technical Bankruptcy Condition 3: $F_t > U_{threshold}\, or \sum_{i=0}^{n} if\left(F_i > U_{threshold,i}, 1{,}0\right) \geq 1$

Fees, once incurred, stay in the system until the system developer releases an update to the system. This update must fix the source of the fees and therefore constitutes a repayment of the original principal and taxes associated with the technical debt. Therefore, repaying the principal can reduce the fees and the associated usage cost.

F_t is expressed as a unitless quantity based on the normalized figures of merit. Technical bankruptcy can occur if the fees from any technical debt item exceed an individual figure of merit threshold. This case occurs when a single compromise has significantly affected the user experience in a specific area. For example, if compromises on mobile device hardware result in the navigation algorithm runtime exceeding five minutes, then the navigation system may not be usable. Technical bankruptcy can also occur if the accumulated fees exceed the overall threshold, $U_{threshold}$. In this case, no individual figure of merit exhibits extremely poor performance, but the accumulated system experience is unacceptable to the user.

Each of these three bankruptcy conditions represents a system state that can lead to technical bankruptcy. The presence of unpaid principal itself is not enough to force a system into technical bankruptcy. Instead, it is the impact of the interest and fees associated with that unpaid principal that leads the system into technical bankruptcy. This relationship is due to the probabilistic nature of technical debt. If the consequence of the technical compromise is never realized, then the system will not enter technical bankruptcy, even if there is a significant amount of technical debt.

12.2.1 Example Quantification of Technical Bankruptcy

To demonstrate how the likelihood of technical bankruptcy can be quantified, consider a situation where the stakeholders for the project specified in Chap. 6.4 are exerting extreme schedule pressure on the system developer to hit the low end of the triangular schedule estimate for every task. This pressure forces the developer to introduce technical compromises. If we assume that the "desired implementation" would have taken the most likely duration for the task, then we can calculate the principal as the difference between these two values.

Let's further assume that there is interest generated by each of these compromises, which affects future tasks and that this interest compounds throughout the system. To estimate the compounding interest, we first assume that interest from a task will increase the duration of any successor task by 30%. If there are multiple predecessor tasks with technical debt, then each one increases the duration of the successor task. The interest factor, f_i, is calculated as shown in Eq. (12.2), where n is the number of predecessor tasks with technical debt.

$$f_i = 1.3^n \tag{12.2}$$

To determine the new expected task duration, d, we multiply the most likely duration, m, times the interest factor using Eq. (12.3).

$$d = m * f_i \tag{12.3}$$

Finally, to isolate the interest contribution, we subtract the original most likely duration from the expected task duration using Eq. (12.4).

$$i = m * f_i - m \tag{12.4}$$

Due to the technical compromises, there will be taxes that need to be repaid when, and if, the technical debt is repaid. The taxes, the amount of work to remove any workaround, are estimated to be half of the interest. The actual time taken, D and the principal, P, for each task in the project are shown beneath the task in Fig. 12.2. For this analysis, we assume that since there is schedule pressure on each task, it will create new technical debt on its successor tasks, and therefore the interest and taxes on each task will vary based on the progress through the development cycle.

In this system, what happens to the balance and potential interest as we progress through the system development cycle? First, group the tasks by predecessors and successors into five different stages of development.

1. Task A completed
2. Tasks B, C, and D completed
3. Tasks E, F, and G completed
4. Task H completed
5. Task I completed.

Define the following values:

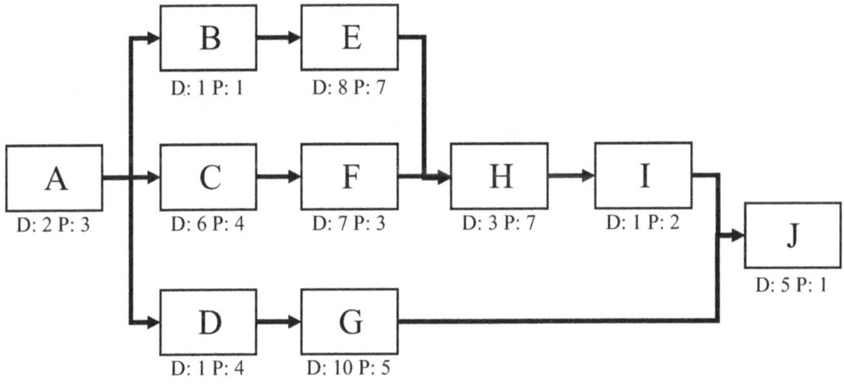

Fig. 12.2 Project schedule for bankruptcy quantification

- The balance, B_t, of technical debt in the system ($P + T$ for all tasks).
- The expected work around cost, $\overline{C_I}$, computed across all unimplemented tasks. If a task has been completed, then we will assume that the interest for that task had to be paid and that we now owe taxes on that task. Therefore, $\overline{C_I}$ is computed as the sum of the interest (I) on unimplemented tasks.
- The difference between the balance and the expected work around cost $\left(B_t - \overline{C_I}\right)$, where negative values indicate the potential for bankruptcy.
- The extra units spent, which counts how much extra work, in the form of paid interest and taxes ($I + T$), needed to be performed to reach this state of the project.
- The margin available at each stage of the project. Assume an initial margin of 60 units. Margin evolves by subtracting the extra units spent in a stage from the margin in the previous stage.

Figure 12.3 shows these values for the situation where no technical debt is repaid throughout the project. Each task creates technical debt during its development, causing interest for its successor tasks. The bars indicate the balance, expected work around cost, and the difference between these two values. The lines indicate the extra units spent at that stage in the project (dashed) and the available margin (solid).

Considering bankruptcy condition 1, $\overline{C_I} > B_t$, Fig. 12.3 shows that the project has the potential for bankruptcy whenever the black column $\left(B_t - \overline{C_I}\right)$ is less than zero. By this rule, the project is potentially bankrupt after completing the first stage of development. This seems awfully early, so let's examine why this happens.

In this case, all tasks in the project are dependent up on the first task (Task A). Therefore, if task A makes a technical compromise, it can compound throughout

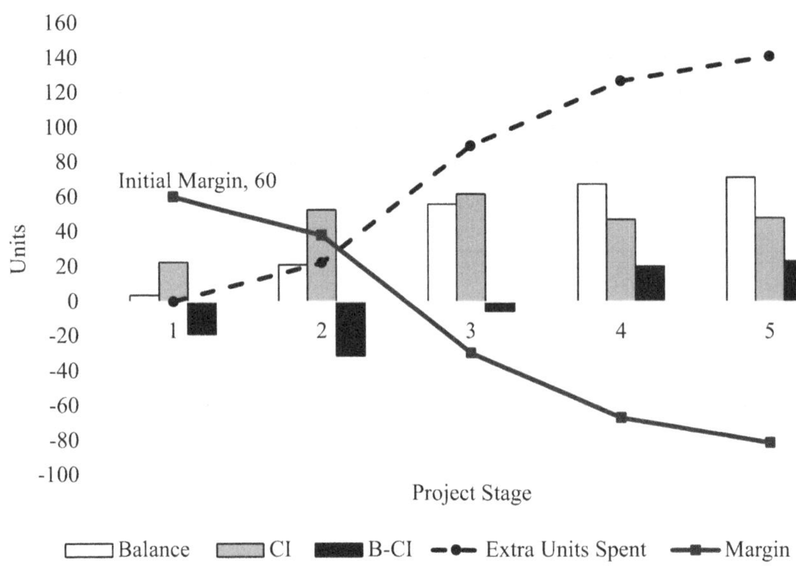

Fig. 12.3 Bankruptcy assessment without managing technical debt

every other task, adding significant interest to the system. Hence, the cost to implement the workarounds for the technical compromise on Task A is larger than the balance of technical debt when only task A is implemented. This situation results from the compounding of technical debt: compromises made on Task A can affect all other tasks in the system, therefore creating larger values of interest than the initial balance. This same rationale applies at stage 2.

However, at stage 3, we see the difference between the balance and the workaround cost decrease, becoming positive in stages 4 and 5 and taking the system out of bankruptcy condition 1. Why is that? In these later stages, the system developer has already paid the interest on many tasks and will have experienced longer development times. The balance is also increasing due to additional principal incurred on each task and the addition of taxes as workarounds are implemented. Therefore, we can expect that the balance will exceed the workaround cost if no technical debt is ever repaid. Bankruptcy condition 1 therefore is a leading indicator of potential technical bankruptcy.

Reading the previous paragraph, you might assume that you don't need to manage your technical debt; after all, we just said that bankruptcy condition 1 no longer applies when the project finishes. However, we have not considered bankruptcy condition 2: $\mathbf{min}\left(\overline{C_I}, B_t\right) > Margin$. This condition states that the system may be in, or heading to, technical bankruptcy if the minimum of the work around costs and the balance is greater than the available margin. In Fig. 12.3, the initial margin is 60 units. As system development proceeds this margin is reduced any time that extra work needs to be completed, as indicated by the dashed line. The solid line shows the resulting margin as each stage is completed.

In this example, the margin becomes less than the minimum of the balance and the work around cost at stage 3, where it also happens that the margin is exhausted. When the margin is used up, then the system is behind schedule or over budget. Therefore, bankruptcy condition 2 is a lagging indicator. It tells us that we are behind and catching up will take more units to repay the balance of the technical debt—not a good spot to be in. Additional resources will be needed. It is not always the case that the margin goes negative when bankruptcy condition 2 is satisfied, but it is the case that recovering from condition 2 will require more resources than the project has remaining.

Figure 12.3 shows the case where the technical debt is not managed. Fig. 12.4 shows two different cases of technical debt management. The left side of the figure shows a case where no technical compromises are made during the first two stages, but then compromises must be made in the following stages. The right side of the figure shows a case where the technical debt from one stage is immediately repaid in the following stage. Therefore, the compounding of the interest is limited.

These approaches have different benefits. In the left case, where technical debt is prevented in the early stages, the overall negative margin is significantly lower at the end of the project–meaning that the project needs fewer resources to reach completion. In the right case, where the technical debt is continuously repaid, the overall number of resources required is similar to the case where no technical management

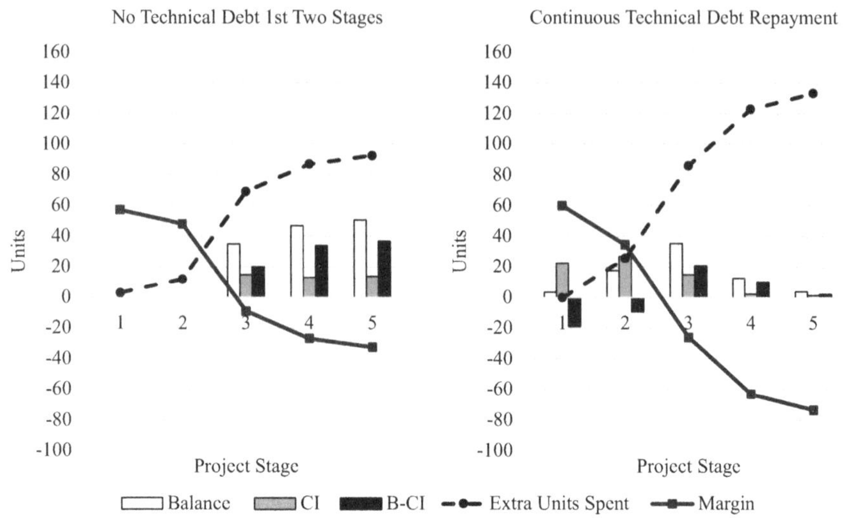

Fig. 12.4 Quantification of technical bankruptcy for with two management strategies: initial avoidance (left) and continuous repayment (right)

occurred (Fig. 12.3). In this case, the additional resources were spent paying down the principal instead of just paying the interest, producing a healthier system at the completion of development. This system, when released to the user, is more likely to be successful than the baseline case.

Technical bankruptcy condition 3 occurs when the fees are greater than the user tolerance for those fees. While fees can be difficult to compute in a generic system, we can estimate the fees based on the balance of the technical debt items. If we assume that each unit of the technical debt balance carries with it 1 unit of fees, then we can plot the fees for each of the three cases assessed above at the time of project release, as shown in Fig. 12.5. Clearly, the case where technical debt is not managed has a higher risk of technical bankruptcy due to the potential for technical fees and the continuous management case provides the lowest fees.

As this example has shown, no one technical bankruptcy condition dominates the other and satisfying a single condition may not be enough to "declare" technical bankruptcy. A system developer should track these conditions and use them as indicators that the system may be facing difficulties in the future. Early indications of technical bankruptcy may be survivable if there is sufficient margin, but may imperil the acceptance of the system. Managing the technical debt early can reduce the total expenditures on the project. However, if margin is used up, late adds of technical debt can still cause excessive fees. The three technical debt conditions, when properly monitored, provide leading indications that a system may be approaching bankruptcy, giving the system developer time to take appropriate actions.

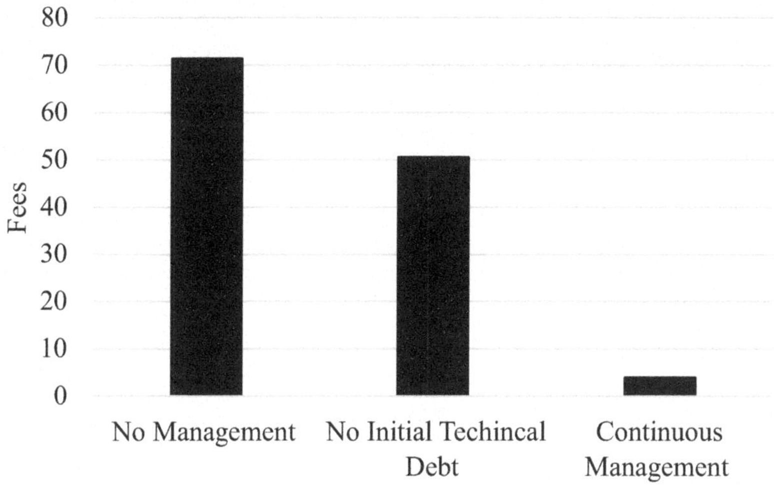

Fig. 12.5 Estimated fees for different management strategies

12.3 Reducing the Risk of Technical Bankruptcy

Technical bankruptcy is a serious condition for any system development. Cost and schedule overruns are serious effects that can trigger formal notice to key stakeholders or even project cancellation. Technical bankruptcy forces cost and schedule increases due to the need to repay principal and taxes beyond the initial project plan. Therefore, reducing the risk of technical bankruptcy must be a key part of any technical debt management program.

The first step in reducing the risk of technical bankruptcy is to ensure that the margins and usage threshold levels ($U_{threshold}$) are known. Establishing these values requires communication with stakeholders (for the margins) and users (for $U_{threshold}$). The developer should evaluate reserves held in the system by the stakeholders to establish the acceptable values for margins. The usage threshold $U_{threshold}$ should be identified for each figure of merit and then the overall threshold computed. Note that the overall threshold is not necessarily the sum of the individual thresholds and should be computed independently.

With these values known, the state of the system against these values can be tracked at the technical debt review board as new technical debt items are added and existing items are mitigated and repaid. Repaying the principal and taxes of a technical debt item has a greater effect on the balance and the fees. As items are repaid, the risk of bankruptcy decreases and the system remains healthy. A technical debt management process is necessary to ensure that the correct technical debt items are prioritized for repayment to reduce the risk of technical bankruptcy.

Another way to reduce the risk of technical bankruptcy is to look into methods to predict where large technical debt items will occur in the development cycle of the system. Everything involved in system development has the potential to create technical debt, including the choice of components in the architecture and the sequence of development. If immature components are selected, they increase the risk of schedule delays during their development which will propagate to the entire system. The development of new capabilities must be phased into a project, especially when adaptive and spiral development strategies are used. The risk of technical bankruptcy is reduced when mature components are incorporated into the final system since the interfaces are known and the performance is established, allowing the design to account for known quantities instead of making assumptions which may never be realized.

12.4 Example of Technical Bankruptcy

To conclude our discussion of technical bankruptcy, let's review an example of a complex program that experienced technical bankruptcy and ended up being cancelled, despite multiple extensions of schedule and budget. Technical debt and technical bankruptcy were not a part of the evaluation of the system, so the values used in this section are my opinions and the quantitative numbers are estimates.

In 1979, the United States Air Force began operating a system called the Space Defense Operations Center (SPADOC) to track objects in space and provide space situational awareness (SSA). SPADOC was upgraded in 1989. By the early 2000s, the rapid pace of technological change drove the Air Force to develop a successor to SPADOC, which was called the Joint Space Operations Center Mission System (JMS) . JMS was designed to replace SPADOC and to update the aging SSA infrastructure which relied on an old mainframe with hardwired data connections. Unfortunately, JMS was cancelled in 2019 due to significant failures during verification and validation [5]. When it was cancelled, JMS was 42% over budget and three years behind schedule [6]. Reviewing JMS through the technical debt lens shows how the program reached multiple instances of technical bankruptcy, which led to its cancellation.

JMS was originally planned to be delivered in three increments: Increment 1 planned to deliver the basic infrastructure, Increment 2 planned to deliver enough capabilities to the user to replace SPADOC, and Increment 3 planned to deliver support for the integration of data from highly classified programs. The program kicked off in 2009 and Increment 2 delivery was planned for the end of fiscal year 2014 [6]. Air Force programs are evaluated by the Air Force Operational Test and Evaluation Center (AFOTEC) and JMS was evaluated in 2016 [7], 2017 [8], and 2018 [9]. Key findings of these results include:

- JMS Increment 2 delivery was late and did not sufficiently test interoperability with other systems, especially Space Fence, a new radar system designed to improve SSA and to supply four times the data compared to existing systems

[5]. The lack of testing increased the risk of successfully interoperating with these systems [7, 8].

- Operational testing was not performed on JMS Service Pack (SP) 7 because it was not intended for mission critical functions, resulting in parts of the deployed JMS system being unvalidated [7].
- In 2018, SP9 was evaluated and the developmental test campaign declared that the delivery was "not operationally effective or suitable for its Space Situational Awareness (SSA) mission" [9] due to numerous critical deficiencies. Due to this test failure, the actual operational delivery was substantially reduced in scope [9].
- Communication between the stakeholders (the program office) and the system developer was a consistent problem throughout the entire system development effort. Improving communication could have prevented many of the problems [9].
- The incremental delivery schedule did not provide time to fix problems observed in one service pack prior to releasing the next service pack. The schedule for SP11 did not provide sufficient time to address the failures of SP9 testing or account for constraints due to parallel development efforts of SP9 and SP11. AFOTEC concluded that the SP11 schedule was not executable [9].

In 2018, JMS was declared not suitable for its mission and not effective for its mission. Consequently, JMS failed its operational test [9], which involved operational users evaluating the performance of the system via its usability, reliability, maintainability, and effectiveness in real-world conditions [10].

12.4.1 Review of JMS Technical Debt

The United States Government Accountability Office (GAO) has produced multiple reports on JMS and we will use these and the AFOTEC reports cited above to assess the evolution of JMS technical debt. Of course, this assessment is easier from our position where we know the results of the program, however, it is still instructive to view how technical debt accumulated within an actual system development and how that system reached technical bankruptcy. An estimated progression of the technical debt within JMS is shown in Fig. 12.6. Major events in the JMS lifecycle are listed to the left of the figure, and the time at which those events occur in the graph is indicated by the corresponding letters. The technical debt balance and the estimated outstanding interest are plotted along with an estimate of the remaining margin, parameters which can be used to evaluate the bankruptcy conditions. Principal was estimated as 1 unit for each technical debt item. Interest was estimated at 1 unit per year for each technical debt item, until the principal was repaid, with the exception of technical debt items A and B, which were estimated at 2 units. Taxes were set at 1 unit per technical debt item, any time interest was paid.

From the start, JMS ran into problems with its development strategy. The program was initially planned to be delivered via five releases over a five-year period from 2011

A. Failure to apply best practices for incremental development and schedule planning
B. Deferral of critical planning activities
C. COTS/GOTS limits to evolvability
D. Not providing planned opportunities for user feedback
E. Reset of program schedule into multiple increments
F. SP7 not operationally tested
G. Schedule breach
H. Project rescope
I. Delayed interoperability testing
J. SP9 scope reduction
K. OT failure

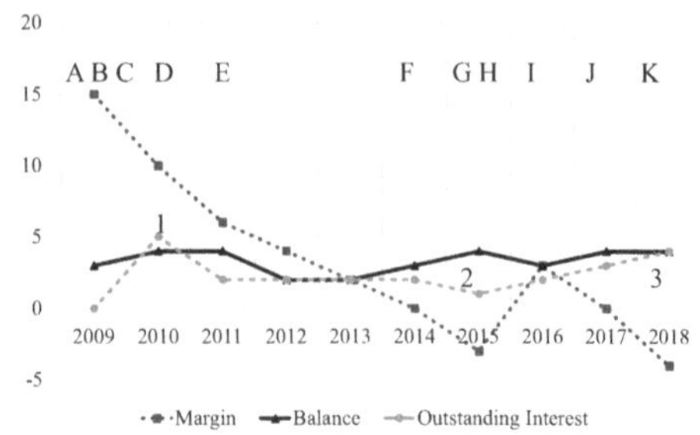

Fig. 12.6 Estimated technical debt within the JMS program, based on GAO reports [7–9, 11, 12]

to 2016. However, the GAO reported that JMS "plans to proceed without knowledge of all critical technologies and deferral of other planning activities" [12], which does not align with the GAO best practices for incremental development (technical debt items A and B). To meet their schedule, JMS made technical compromises and incurred design debt by beginning system implementation before the design was completed. Dependencies between components and capabilities were not fully defined and therefore the interest associated with the technical debt caused rework of the delivered system to adjust to the final interface definitions. In some cases, the technologies that would be used in future increments were undefined, meaning that the initial increments ran the risk of designing out critical capabilities.

Early in the program, the JMS development team decided to use only existing software, including both commercial-off-the-shelf (COTS) and Government-off-the-shelf (GOTS) products [11]. However, these existing software components were insufficient to meet the capabilities of the system, requiring significant work (and increases in cost and schedule) to enable the components to work together. These software components also locked the JMS system into designs due to the limitations of the software components, preventing JMS from adapting to changes to meet user needs. These issues can be seen in the delay of testing with the Space Fence, where software interoperability issues required increased efforts to enable communication between the different systems. If the JMS program office had reviewed the requirements more carefully, they could have chosen COTS software that better met their needs or would have included additional time in the schedule to alter the selected software. The selection of insufficient COTS software introduced significant infrastructure debt (technical debt item C).

The JMS program also incurred domain debt due to its failure to engage with users early in the development process, even though early user engagement was part of the initial plan (technical debt item D). Users were not consulted during the design process nor were they given opportunities to review demonstrations and provide feedback on the software as it was developed [11]. This domain debt is an example of reckless and deliberate technical debt–the JMS developers knew that they should include users (it was in the plan) but chose not to do so, resulting in a system that did not align with the users' needs.

12.4.2 Attempts to Repay JMS Technical Debt

We can look at how the program restructured through the information provided in the GAO reports and infer how the JMS program attempted to repay its technical debt. In 2011, GAO recommended that the JMS program restructure its increment-based delivery strategy into multiple increments and JMS implemented that recommendation (event E) [11]. This restructuring repaid the balance associated with technical debt item A (proceeding with the design without a complete knowledge of the architecture). Replanning the schedule to cover multiple increments enabled better design in each increment.

There is little evidence of other activities that JMS took to repay technical debt. What we do see are indications of the interest being paid through failed tests and delays in integration. The infrastructure debt associated with technical debt item C (insufficient COTS/GOTS software) led to the failure of some of the operational tests. The operational test evaluation, which declared JMS "not operationally effective or suitable for its Space Situational Awareness (SSA) mission" [9], highlighted that JMS had "underperforming mission functions, system instability," and "configuration problems," which are all symptoms of realized infrastructure debt. Interoperability testing with the Space Fence, a major new SSA sensor, was delayed and was more complicated than expected, representing interest payments on the design debt. SP7 was developed but not operationally tested due to its lack of mission critical functions (technical debt item F). This lack of operationally relevant testing constitutes test debt–even though the service pack did not include critical functions, getting test results on the software when it was completed would have ensured that it was properly developed. Additionally, the failure to include mission critical functions is an example of interest on the domain debt: extra work to develop non-mission critical capability represents a lack of knowledge of the application domain.

12.4.3 JMS Technical Bankruptcy

There were three potential instances of technical bankruptcy in the JMS program (labelled 1, 2, and 3 in Fig. 12.6). The first occurrence was in 2010 and occurred when the outstanding interest was greater than the technical debt balance, satisfying bankruptcy condition 1. This occurrence is estimated to have occurred due to the fact that the JMS program did not follow schedule best practices and had not met with the users as planned (technical debt items A and D). The schedule was realigned into increments in 2011, which paid down some of the existing balance and moved the system out of bankruptcy.

The second instance of technical bankruptcy occurred in 2015 when JMS declared a schedule breach (event G) when its schedule exceeded its initial milestones [11]. The GAO assessments found that JMS would be unable to meet the planned schedule delivery dates and milestones [13]. This breach satisfied technical bankruptcy condition 2: the available margin was less than the minimum of the balance and the outstanding interest. JMS resolved this bankruptcy by rescoping the goals (event H) and obtaining schedule relief from their stakeholders, which added margin back into the program. Without this relief, the program would not have been able to continue in its lifecycle. However, additional technical debt was added to the system with further delays to interoperability testing (technical debt item I) and the SP9 scope reduction (technical debt item J). The third and final technical bankruptcy for JMS occurred at the end of the program and satisfied technical bankruptcy conditions 2 and 3: it was out of margin and the fees had exceeded the user thresholds, as evidenced by the JMS system failing its operational test (event K) [9]. These repeated technical bankruptcies contributed to program cancellation.

12.5 Summary

In this chapter, we provided a detailed review of technical bankruptcy. We developed conditions to quantitatively identify technical bankruptcy based on the measured technical debt in the system and stakeholder thresholds. These conditions are:

- The expected work around cost exceeds the technical debt balance;
- The minimum of the balance and the work around cost is greater than the program margin; and,
- The total fees are greater than the user's threshold.

By reviewing these conditions regularly, the system developer can take actions prior to reaching the state of technical bankruptcy, thus minimizing risks of the system lifecycle being stopped. In the final chapter, we conclude the book with some strategies for implementing a technical debt management program in your organization.

References

1. Ciolkowski, M., Lenarduzzi, V., & Martini, A. (2021). 10 Years of Technical Debt Research and Practice: Past, Present, and Future. *IEEE Software, 38*(6), 24–29. https://doi.org/10.1109/MS.2021.3105625
2. Li, Z., Avgeriou, P., & Liang, P. (2015). A systematic mapping study on technical debt and its management. *The Journal of Systems and Software, 101*, 193–220. https://doi.org/10.1016/j.jss.2014.12.0-27
3. Defense Acquisition University. (n.d.). *Nunn-McCurdy Breach*. Defense Acquisition University. Retrieved May 5, 2025, from https://www.dau.edu/glossary/nunn-mccurdy-breach
4. United States Government Accountability Office. (2011). *DOD Cost Overruns: Trends in Nunn-McCurdy Breaches and Tools to Manage Weapon Systems Acquisition Costs*. (GAO-11-499T).

5. Clark, C. (2019, April 8). *What About JMS? Air Force Reanimates 'Old Clunker' Space Tracking System.* Breaking Defense. Retrieved May 10, 2025, from https://breakingdefense.com/2019/04/what-about-jms-air-force-reanimates-old-clunker-space-tracking-system/
6. United States Department of Defense, Office of the Under Secretary of Defense for Acquisition and Sustainment. (2021). *DoD Instruction 5000.85: Major Capability Acquisition.*
7. Gilmore, J. M. (2016). *Director, Operational Test and Evaluation FY 2016 Annual Report.* The Office of the Director, Operational Test and Evaluation. Retrieved from https://www.dote.osd.mil/Portals/97/pub/reports/FY2016/other/2016DOTEAnnualReport.pdf?ver=2019-08-22-105134-547
8. Behler, R. F. (2017). *Director, Operational Test and Evaluation FY 2017 Annual Report.* The Office of the Director, Operational Test and Evaluation. Retrieved from https://www.dote.osd.mil/Portals/97/pub/reports/FY2017/other/2017DOTEAnnualReport.pdf?ver=2019-08-19-113553-787
9. Behler, R. F. (2018). *Director, Operational Test and Evaluation FY 2018 Annual Report.* The Office of the Director, Operational Test and Evaluation. Retrieved from https://www.dote.osd.mil/Portals/97/pub/reports/FY2018/other/2018DOTEAnnualReport.pdf?ver=2019-08-21-155621-540
10. Air Force Operational Test and Evaluation Center. (n.d.). *Air Force Operational Test and Evaluation Center.* Retrieved July 9, 2022, from https://www.afotec.af.mil/About-Us/Fact-Sheets/Display/Article/872935/air-force-operational-test-and-evaluation-center/
11. United States Government Accountability Office. (2019). *DoD Space Acquisitions: Including Users Early and Often in Software Development Could Benefit Programs.* (GAO-19-136).
12. United States Government Accountability Office. (2011). *Space Acquisitions: Development and Oversight Challenges in Delivering Improved Space Situational Awareness Capabilities.* (GAO-11-545)
13. United States Government Accountability Office. (2015). *Defense Major Automated Information Systems: Cost and Schedule Commitments Need to Be Established Earlier.* (GAO-15-282).

Chapter 13
Conclusion

Abstract Technical debt management may be a new concept to many system developers. When done properly, it enables the system developer to produce short-term benefits while limiting the long-term consequences of decisions made to satisfy stakeholders. The development of the Deep Space Network by the Jet Propulsion Laboratory is provided as an example of successful technical debt management. The first steps to implementing a technical debt management process in an organization are provided. These steps, which can be tailored based on the organization, are a guide to starting to manage technical debt and to prevent technical bankruptcy.

Keywords Implementing technical debt management program · Successful technical debt management

That's it! We've reached the end of the book. We started with the origins of the technical debt metaphor, developed practical definitions for technical debt in the context of systems engineering, established strategies for managing technical debt during systems development, and quantified technical bankruptcy. We've covered a lot of ground, and it may seem like every system is destined for technical bankruptcy or that the technical debt management process will only cause delays. The good news is this is not true!

Like any process change, building technical debt management into your system development processes will take time. Multiple iterations may be required to successfully tailor the recommendations in this book to your organization. In this chapter, we first review a successful case of technical debt management and then conclude the book with some strategies for beginning a technical debt management program.

H. Kleinwaks, *Managing Technical Debt in Systems Engineering*,
https://doi.org/10.1007/978-3-032-30360-8_13

13.1 A Case Study in Successful Technical Debt Management

Most of the examples of technical debt that we provided so far focused on unsuccessful technical debt management. We looked at situations where technical debt, whether it was identified by the system developer or not, grew until it caused major system failures, delays, or cost overruns. However, there are successful examples of technical debt management, even from the time before the metaphor was introduced. In this section, we will look at one such case because it directly compares two competing companies and shows how limiting technical compromises and preserving the long-term health of the system resulted in one company being successful and one company failing.

In his book on technical road mapping, Olivier de Weck discusses the evolution of the Deep Space Network (DSN) , the set of ground antennas used to communicate with interplanetary spacecraft [1]. Two organizations, the Jet Propulsion Laboratory (JPL) and the Space Technology Laboratory (STL), received contracts to launch the Pioneer spacecraft to the moon, inclusive of the development of a communications architecture. de Weck frames the discussions around the successful decisions that JPL made to create the modern DSN. However, looking at the same situation from a technical debt perspective reveals that JPL avoided technical compromises that hurt the long-term health of the system, while STL made many technical compromises to expedite their system development whose consequences could not be overcome. We've adapted the following discussion from the Ph.D. dissertation of Jennifer Manuse [2].

In March of 1958, the United States (US) government was still feeling negative political pressure due to the successful Soviet launch of Sputnik in October 1957. The first US satellite launch attempt, Vanguard, exploded on the launch pad on live television, resulting in embarrassment and reinforced public views that the US was significantly trailing in the space race. Although the US did successfully launch the Explorer I mission in January 1958 and Vanguard 1 in March 1958, the US still felt the need to advance in the space race.

In March 1958, the US government approved funding for the Pioneer program—a one-year program to launch five lunar probes with a mission to make "significant contact" with the moon, such as by returning pictures of the lunar far side [3]. Three probes were awarded to the Air Force and STL and two probes were awarded to the Army and JPL. This program required developing new technology in three areas: spacecraft, launch vehicles, and ground systems. The ground segment of the Pioneer program needed to be able to command and control the spacecraft and receive telemetry data from the spacecraft, or else the program would be unsuccessful.

In developing their ground segments, STL and JPL took different approaches. STL, which was developing the first three Pioneer probes, focused exclusively on meeting the needs of the Pioneer program, due in part to the limited five-month timeline to set up the network. This schedule pressure forced technical compromises. They only considered existing sites for their ground antennas, which limited them to existing hardware and capabilities. Political pressure required that the antenna be in the territory of an allied nation. Technical trade-offs preserved the orbital geometry of the mission but forced the antenna to be located near the prime meridian. With all of these constraints, STL selected a 250-ft diameter radio telescope in Jodrell Bank, England and installed temporary equipment to enable communications with the probe, using the already assigned frequencies. Engineers at STL realized that a more permanent solution would be required, indicating that STL knew they were making a technical compromise. However, STL did not take an active role in developing the network.

JPL took a different approach. They had an additional three months (eight months total) before the launch of their first probe and decided to build their own ground station network. Under the leadership of Eberhardt Rechtin, they designed a purpose-built set of antennas to enable not only the Pioneer missions but also future missions. JPL had similar schedule pressures and political pressures as STL; however, they decided that they could proceed without taking on the technical debt. If they took on the technical debt, they would be risking the ability to complete the future missions, which would impact the long-term health of the system they wanted to build. They selected to manufacture a new antenna and designed the mission operations schedule to enable communications from a location within the US, reducing the diplomatic hurdles associated with foreign sites. Furthermore, since they were designing new antennas, they selected a frequency that enabled higher data rates and therefore supported future growth of that solution.

While both ground systems performed properly, only JPL was able to expand the ground network following the (mostly unsuccessful) Pioneer launches. STL proposed expanding their network with additional antennas similar to the one in Jodrell Bank. However, while the JPL antennas cost $250,000, the STL antenna was estimated at $34 million and was not expected to meet the proposed timeline. JPL's future expansions, with cheaper antennas and expanded bandwidth, were selected for future development and eventually became today's Deep Space Network.

It is easy to look at this situation as a success of JPL management strategy or as a failure of STL management strategy. We can also apply the lens of technical debt to the scenario to determine where JPL succeeded and STL failed. STL introduced significant architecture debt into their system through the selection and modification of the Jodrell Bank radio telescope. This selection represents an instance of selecting an MVP for a specific mission (architecture debt: MVP that stuck). This selection limited the ability of the system to expand, as evidenced by the high cost for a second antenna (architecture debt: new context, old architecture). JPL, on the other hand, developed an architecture that worked for both current and future missions, limiting the long-term consequences. JPL focused on designing the entire system, including modifying the frequencies, to build an enduring product. By taking a systems-level approach, JPL identified technical trade-offs to support the Pioneer mission and to position themselves to support future missions. JPL had less schedule pressure than STL (an additional three months), but their willingness to examine multiple parameter sets in the architecture limited the growth of technical debt and enabled the continuing success of the DSN today.

13.2 First Steps to Implementing a Technical Debt Management Program

With our discussion of technical debt in systems engineering complete, it is now up to you to implement these strategies in your organization to help minimize the impacts of technical debt. While this may seem like a daunting task, there are a few simple steps that can be taken to establish the new processes and procedures necessary to make technical debt management part of the standard workflow.

1. Establish the lexicon: The first step in technical debt management is talking about technical debt management. It is necessary to start introducing the terminology and the ontology into standard discussions to bring awareness of the problems to the forefront. If you start talking about technical debt and looking for examples of technical debt in your system, then others will start talking about technical debt as well. Once people are talking about it, and recognizing that it has impacts on your system, then you'll be able to start doing something about it.
2. Start a technical debt register: Introducing the lexicon is not enough if it is not backed by data. Using some of the techniques from Chap. 8, find existing technical debt items in the system. This step is not designed to cast blame on people for introducing technical debt, but rather to show that there could be lurking problems within the system that need to be addressed. For each identified item, specify the principal, interest, taxes, and fees and their potential long-term negative impact on the system health. Log the technical debt items into a technical debt register. Identify a mitigation plan to minimize the impacts from these technical debt items.
3. Get leadership buy in: With the lexicon established and the start of a technical debt register, make your case to leadership for why a technical debt management program is necessary. Show how you can identify issues and mitigate them before the cost (in time or money) to do so becomes excessive. Leadership support is essential to making the process changes associated with technical debt management a permanent part of the system development.
4. Include technical debt assessments in the existing review processes: Leverage your existing system reviews (major gate reviews, peer reviews, ERBs, ROMBs, etc.) to start discussing potential technical debt in the system. At each review, explicitly address the technical debt associated with the system (or portion of the system) under review. This step helps to populate the technical debt register and reinforces the need to identify technical debt across the entire workforce.
5. Establish a technical debt review board: Add a separate technical debt review board to your standard review boards to prioritize the technical debt register for repayment and to validate the identified instances of technical debt. This board owns the technical debt processes and can continue to tailor and modify them to best fit your organization.

Following these steps will help establish an initial technical debt management process, which can evolve as needed within your organization. I have introduced such a process at my organization and now technical debt is considered as part of every major decision and communicated across teams. Operations team members assess the future consequences of the design decisions made by development teams to identify potential long-term impacts. This information sharing has reduced risks associated with the future spirals by bringing awareness of issues and allowed those future spirals to adapt designs as needed to prevent the spread of the technical debt. By tracking our technical debt, our teams working on the early stages of future spirals are able to understand the restrictions and additional work placed on them through the acceptance of technical debt in current spirals.

13.3 Conclusion

As we conclude this book, it is important to review the major topics that we covered. The concept of technical debt, well established in software, is not as well established in the systems engineering community. However, as systems engineering becomes more agile and more dependent upon software systems, the impacts of technical debt are readily apparent in systems development. Therefore, it is essential to consider technical debt management throughout the development of any system, not just in software systems.

Technical debt can appear throughout the systems engineering lifecycle and occurs when pressures, usually from stakeholders, force the system developer to make a technical compromise. This compromise achieves a short-term benefit but also introduces the potential for negative long-term impacts on the health of the system. Technical debt can appear in any system lifecycle, but the added schedule pressures of incremental and adaptive methods make it more prevalent in these systems.

While we know that technical debt exists in systems engineering, there is not a consistent ontology in use. We presented such an ontology in Chap. 3. Without the ability to express common ideas in common language, we cannot hope to come together on developing management strategies. In addition, we covered common types and causes of technical debt, introducing these as starting points for organizations. It is likely that each organization and each project will encounter their own specific types and causes and will want to tailor the starting list that we provided.

Finally, we discussed the technical debt management cycle in detail, starting with methods for identifying technical debt and moving through the measurement, prioritization, repayment, and monitoring stages. Each of these stages plays a critical role in minimizing the risk of technical bankruptcy in a system and enabling it to be a continued success.

Technical debt management is not intended to force system developers to build a system that can handle any possible outcome. Instead, it is about looking at the desired end state of the system and asking yourself, "If I make this decision, will it make it easier or harder to get there?" Sometimes, what appears to be the easier or better path (the short-term benefit) will actually result in a longer and more expensive journey. Technical debt left in the system can show its impacts at the most inopportune time. However, proper identification and management of technical debt will limit the impacts of these latent defects and minimize the risk to a successful system development.

I hope that you have enjoyed this book and learned something from it. I urge you to consider the potential for technical debt in your decisions moving forward. Remember, not all technical debt is bad, you just need to have a plan to manage it.

References

1. de Weck, O. L. (2022). *Technology Roadmapping and Development: A Quantitative Approach to the Management of Technology*. Springer Nature.
2. Manuse, J. E. (2009). *The strategic evolution of systems: principles and framework with applications to space communication networks*. Doctoral Dissertation, Massachusetts Institute of Technology, Department of Aeronautics and Astronautics.
3. Waff, C. B. (1993, April). The road to the Deep Space Network. *IEEE Spectrum, 30*(4), 50–57. https://doi.org/10.1109/6.206623

Appendix A
Including Technical Debt in Schedule Analysis

Abstract Traditional schedule risk assessments vary the duration of each task but assume that possible durations of subsequent tasks are unaffected by the quality of completion of predecessor tasks. Technical debt often manifests as increases in the duration of subsequent tasks. A method is presented to include probabilistic increases of the duration of subsequent tasks based on the completion of predecessor tasks. This method shows the impacts of technical debt within a schedule and how it compounds throughout the system development if left unmanaged.

Keywords Technical debt in schedule risk assessments · Monte Carlo analysis

Monte Carlo schedule analysis adds uncertainty to the duration of each task; however, it often overlooks the compounding effects of technical debt on task durations. Interest-bearing technical debt increases the effort required to develop future components of the system. Therefore, introducing technical debt in one component does not affect the duration of that component, but instead the technical debt affects the duration of the succeeding components. Traditional Monte Carlo analysis does not account for this impact—the duration probability for each task remains constant regardless of the quality of the predecessor tasks.

Therefore, a method is required to establish probability distributions for the impacts of technical debt on the duration of successor tasks such that they can be used within the Monte Carlo analysis. When implemented, components with high risks for technical debt will not see their duration increase outside of the initial estimates. Downstream components that depend on these debt-ridden components will see significant increases in their durations, including going beyond the durations specified by the initial duration parameters.

H. Kleinwaks, *Managing Technical Debt in Systems Engineering*,
https://doi.org/10.1007/978-3-032-30360-8

To include technical debt in a Monte Carlo schedule analysis, we must add a multiplier to the duration of each task to adjust the task duration based on the technical debt from the predecessor tasks. These multipliers would be determined from a probability distribution, which is used to represent the interest. Within a single run of the Monte Carlo analysis, the duration, D, of a given task t is calculated using Eq. (A.1).

$$D_t = d_t \prod_{i=0}^{p} m_{t,i} \tag{A.1}$$

In this equation, d is the planned duration of the task, m is the multiplier to the duration due to technical debt from predecessor task i, and p is the total number of predecessor tasks. The values for d and m would be generated according to the specified probability distributions, with m always being greater than or equal to one (1) such that a predecessor task with no technical debt would produce a value of one (1) and not affect the duration of the successor task. This approach assumes that a task does not start until its predecessor completes. Additional techniques can be generated to explore the impacts of parallel tasking in the presence of technical debt, such as starting code development prior to the completion of an ICD.

To examine the impact of technical debt, let's start by estimating the duration of a single task. A triangular distribution is a simple approach used in schedule analysis to estimate task duration. In a triangular distribution, the low, medium, and high values are specified, representing the shortest possible duration (low value, L), most likely duration (medium value, M), and the longest possible duration (high value, H). The resulting distribution is shaped like a triangle, with the high point (and most frequent outcome) equal to the specified medium value. Figure A.1 shows a triangular distribution with values L = 3, M = 7, and H = 9. This distribution is often used because the distribution parameters are intuitive—experts can use experience to determine the expected duration (M), and the best (L) and worst (H) cases.

Now assume that the task has a single predecessor. We can use the same triangular distribution to estimate the multiplier on the task duration due to technical debt from its predecessor. There is a chance that the predecessor will not impart any technical debt onto the task, so we set the low value in the distribution to 1. We then need to estimate the impacts of any technical debt interest. We assume that the worst-case scenario results in doubling the length of the task and that the most likely case is that the task takes 30% longer than expected. The resulting distribution of m is shown in Fig. A.2.

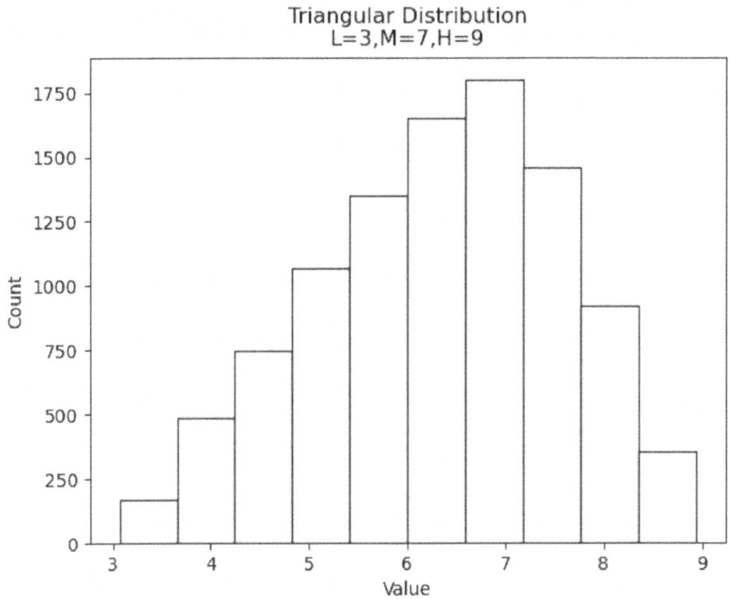

Fig. A.1 Triangular distribution

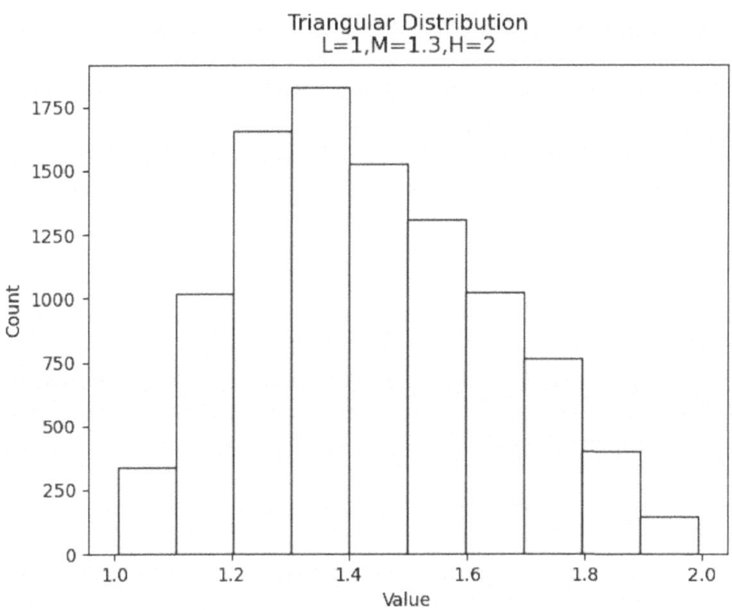

Fig. A.2 Triangular distribution of technical debt multiplier, m

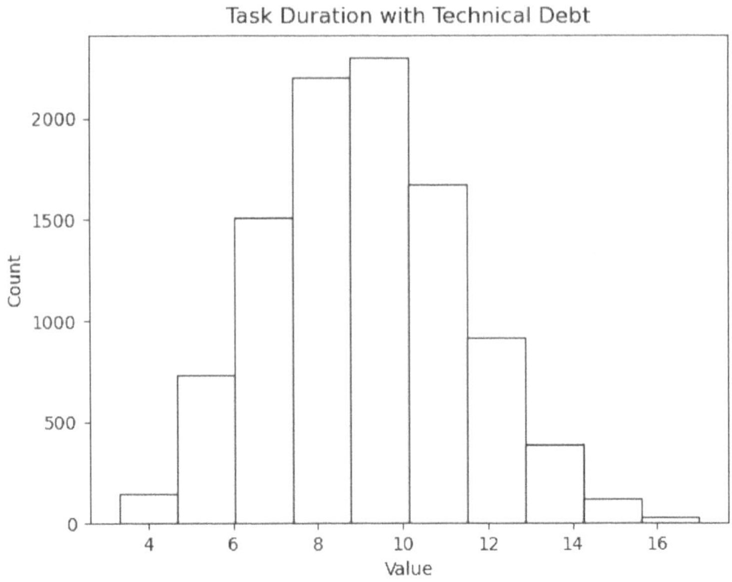

Fig. A.3 Distribution of task duration with technical debt

With each probability distribution determined, we can now specify the combined distribution for the duration of the task, accounting for the technical debt, shown in Fig. A.3. In this distribution, the task duration centers on nine (9), which is a shift from the unaffected task duration, which centered on seven (7).

Using this method, we can examine the effect on a simple project schedule to show how technical debt interest can build up in a project and compound its effect on tasks near the end of the development cycle. Consider the notional project shown in Fig. A.4. This project consists of ten tasks, and has three separate, parallel development paths that each begin after the completion of the first task (Task A). Below each task are the triangular distribution parameters for the task, where L is the low estimate, M is the medium value, and H is the high value, all of which are specified in days. This project consists solely of finish-to-start dependencies, meaning that the successor tasks cannot complete until the predecessor tasks have completely finished.

There are three paths through this project: (1) A-> B-> E-> H-> I-> J, (2) A-> C-> F-> H-> I-> J, and (3) A-> D-> G-> J. If we were to only look at the mostly likely durations, then the duration for path 1 is 40 days, the duration for path 2 is 43 days, and the duration for path 3 is 30 days. The critical path, the continuous string of activities between the project start and finish that determines the project duration, is path 2, and the overall project duration is 43 days. Of course, we need to account for random factors that change the durations and execute the Monte Carlo analysis.

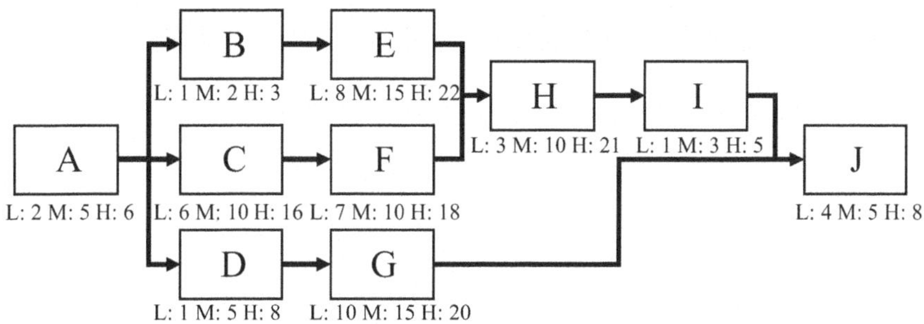

Fig. A.4 Notional project

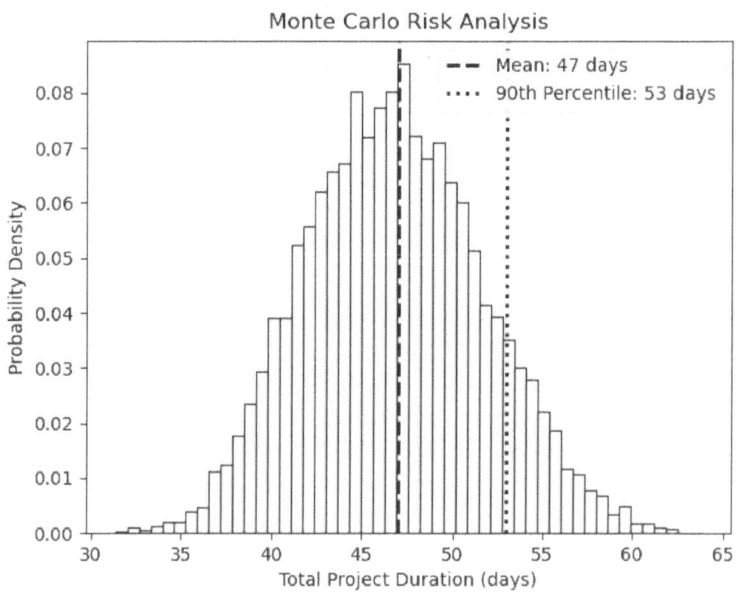

Fig. A.5 Baseline Monte Carlo schedule analysis

Using Python code generated by Microsoft Copilot and then manually modified, a baseline Monte Carlo schedule run produces a mean project duration of 47 days and a 90th percentile of 53 days, meaning that the 90% of the time the project will complete in less than 54 days. The schedule distribution is shown in Fig. A.5. We can also examine the critical path in each trial—89% of the time, the critical path is path 2. However, 11% of the time, the critical path is path 1. For this to happen, everything needs to go right on path 2 while everything goes wrong on path 1.

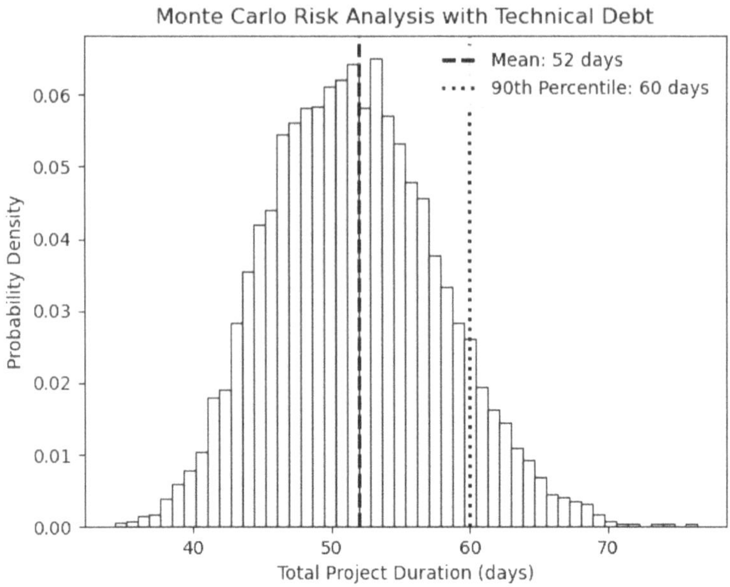

Fig. A.6 Monte Carlo schedule risk analysis with technical debt on Task C

With this baseline understanding of the project schedule, we can start to examine the impacts of technical debt. First, let's look at the impacts of technical debt on a single task. Task C has the potential to be a long-duration task, with a high estimate of 16 days. This task is also on the critical path, indicating that there will be a day-for-day slip in project delivery if Task C is delayed. Therefore, there is likely to be schedule pressure to complete Task C on time, and schedule pressure can create technical debt.

To model this technical debt, we will assume that the duration estimates of Task C are unchanged, but that the accrued technical debt affects the duration of the immediate successor task, Task F. Using the method outlined above, set the triangular distribution for the multiplier to L = 1, M = 1.3, H = 2, meaning that the worst-case outcome is a doubling in the duration of Task F and the most likely outcome is a 30% increase in its duration. The results of the updated Monte Carlo analysis are shown in Fig. A.6, where it can be clearly seen that the average project duration increased to 52 days and the 90th percentile value increased to 60 days.

Figure A.7 shows the average durations of each task across the project, both with and without the technical debt from Task A. The duration of Task C is unaffected, but the average duration of Task F increased by 5 days, which accounts for the overall increase in the project schedule.

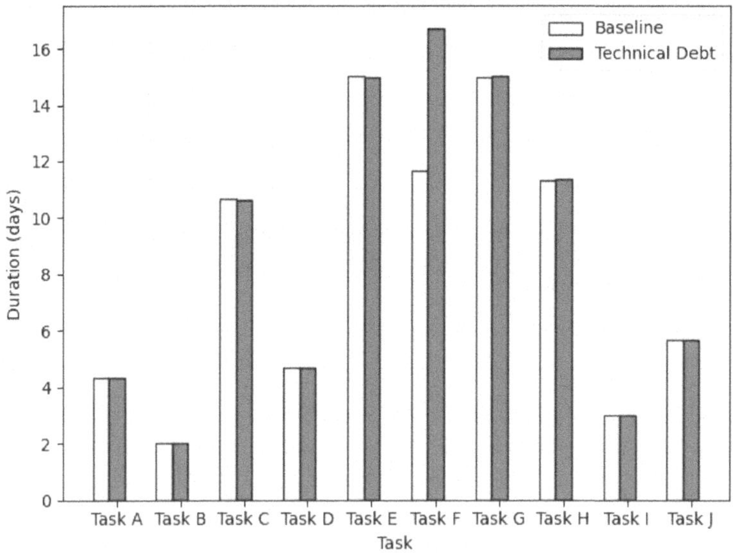

Fig. A.7 Individual task duration due to technical debt on Task C

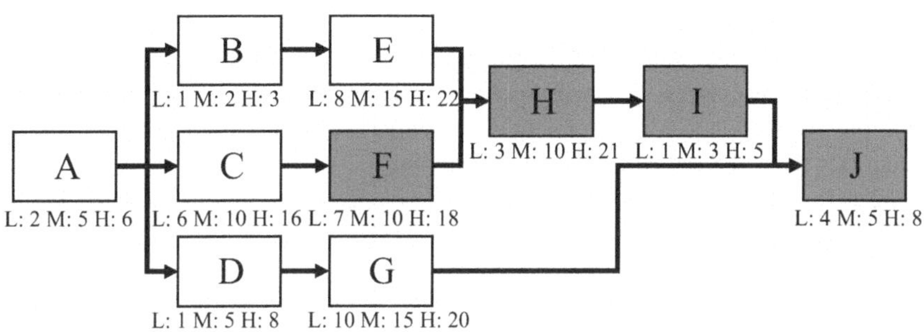

Fig. A.8 Tasks affected by compounding technical debt on task C

This example shows a simple case of technical debt—one task only affects its successor task. However, technical debt interest compounds. The technical debt accrued with one technical compromise may affect several tasks within the system, and these tasks may be significantly farther downstream in the development cycle. We will now model these effects.

To model the compounding technical debt, consider the case where the technical debt accrued on Task C affects every successor task, and not just its immediate successor task. For simplicity, start with the same distribution for the technical debt multiplier on each successor task: L = 1, M = 1.3, H = 2. The affected tasks are shown in gray in Fig. A.8.

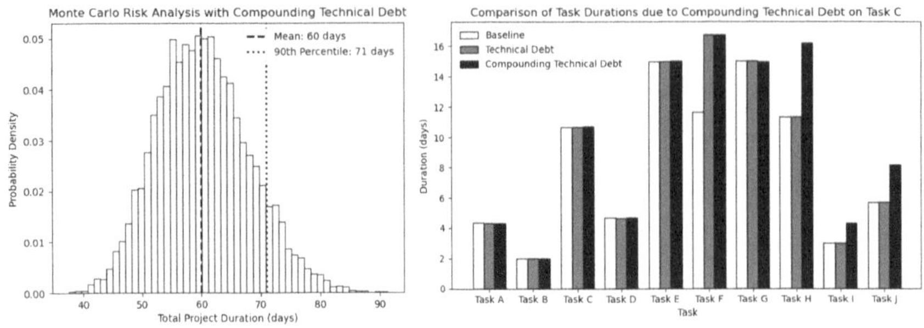

Fig. A.9 Monte Carlo schedule analysis with compounding technical debt

With the increased number of affected tasks, we would assume that there would be an overall increase in the project schedule. This is indeed the result, as shown in Fig. A.9. The impacts of the technical debt from Task C have increased the average project duration to 60 days, which represents a 15% increase over the schedule where only Task F was affected by the technical debt and a 28% increase over the original baseline schedule. Similarly, the 90th percentile duration is now up to 71 days. Figure A.9 also shows the effect on the individual tasks—Task H, I, and J all show increased durations due to the technical debt from Task C.

We could continue to run multiple variations of technical debt, modeling its effects on different tasks to show which have the greatest inputs, and also varying the interest amount and probabilities. This work is beyond the scope of this book; however, we encourage the reader to add technical debt considerations into their project schedule estimations to see how it can impact the schedule.

The critical aspect of including technical debt in schedule risks analysis is that even if a task finishes on time, it may not be at the expected quality level, thus imparting technical debt onto its successor tasks. Therefore, we need to include modifiers within the analyses to account for these impacts to more realistically evaluate the schedules in the presence of technical debt.

The Python code used to perform the schedule analysis follows.

```
# -*- coding: utf-8 -*-
"""
Created on Sun Apr 6 09:49:14 2025
@author: hkleinwaks, with an assist from Microsoft Co-
pilot
"""
import numpy as np
import matplotlib.pyplot as plt

def monteCarloAnalysis(tasks, num_simulations, title):
  #generated by Microsoft Copilot with the following
prompt (4/6/2025):
  #"write python code to run a monte carlo schedule risk
analysis with parallel tasks and conditional probabili-
ties"
   #Define tasks with optimistic, most likely, pessimis-
tic durations (in days)
  #Include dependencies and conditional probabilities
  # num_simulations = 10000 # Number of Monte Carlo it-
erations
  total_durations = []
  taskDurations = {}
  for _ in range(num_simulations):
    start_times = {}
    completion_times = {}
    for task, details in tasks.items():
      # Default durations
      optimistic, likely, pessimistic = details["dura-
tion"]
       # Apply conditional probabilities if a condition
is defined
      if details["condition"]:
        dependent_tasks = details["condition"]["depend-
ent_on"]
        interest_amounts = details["condition"]["inter-
est_amount"]
        #get the multiplier from the distribution
        #probability that we see interest
        multiplier = 1
        for i in range(0, len(dependent_tasks)):
          multiplier*=np.random.triangular(inter-
est_amounts[i][0],    interest_amounts[i][1],    inter-
est_amounts[i][2])
        optimistic *= multiplier
        likely *= multiplier
        pessimistic*= multiplier
      # Simulate task duration using triangular distri-
bution
```

```
      duration    =    np.random.triangular(optimistic,
likely, pessimistic)
      # Determine start time based on dependencies
      if  not  details["dependencies"]:  #  Independent
tasks start at 0
        start_times[task] = 0
      else: # Start after the latest completion of de-
pendent tasks
        start_times[task]  =  max(completion_times[dep]
for dep in details["dependencies"])
      # Calculate completion time
      completion_times[task] = start_times[task] + du-
ration
      if (not (task in taskDurations)):
        taskDurations[task] = []
      taskDurations[task].append(duration)
    # Total project duration is the completion time of
the last task
    total_durations.append(completion_times["Task J"])
  # Compute statistics
  mean_duration     =     int(round(np.mean(total_dura-
tions),0))
  percentile_90  =  int(round(np.percentile(total_dura-
tions, 90),0))
  # Plot histogram of results
  fig, ax = plt.subplots()
  plt.hist(total_durations, bins=50, density=True, al-
pha=0.75, color='white', edgecolor='black')
  plt.axvline(mean_duration,    color='black',    lin-
estyle='dashed', linewidth=2, label=f"Mean: {mean_dura-
tion} days")
  plt.axvline(percentile_90,    color='black',    lin-
estyle='dotted',  linewidth=2,  label=f"90th  Percentile:
{percentile_90} days")
  plt.xlabel("Total Project Duration (days)")
  plt.ylabel("Probability Density")
  plt.title(title)
  plt.legend()
  plt.show()
  averages = []
  #print("average completion time of each task")
  for task, details in tasks.items():
    #print(task, np.average(completion_times[task]))
    averages.append(np.average(completion_times[task]))
  averageDurations = []
  print("average duration of each task")
  for task, details in tasks.items():
```

```
    print(task, np.average(taskDurations[task]))
    averageDurations.append(np.average(taskDura-
tions[task]))
  print(f"Expected project duration: {mean_duration:.2f}
days")
  print(f"90th percentile duration: {percentile_90:.2f}
days")
  return averageDurations
#execute the code
if __name__ == "__main__":
  n = 10000 #run 10000 simulations
  #set up a baseline project schedule with 10 tasks and
3 execution paths, per the figure
  tasks = {
    "Task A": {"duration": (2, 5, 6), "dependencies":
[], "condition": None},
    "Task B": {"duration": (1, 2, 3), "dependencies":
["Task A"], "condition": None},
    "Task C": {"duration": (6, 10, 16), "dependencies":
["Task A"], "condition": None},
    "Task D": {"duration": (1, 5, 8), "dependencies":
["Task A"], "condition": None},
    "Task E": {"duration": (8, 15, 22), "dependencies":
["Task B"], "condition": None},
    "Task F": {"duration": (7, 10, 18), "dependencies":
["Task C"], "condition": None},
    "Task G": {"duration": (10, 15, 20), "dependencies":
["Task D"], "condition": None},
    "Task H": {"duration": (3, 10, 21), "dependencies":
["Task E", "Task F"], "condition": None},
    "Task I": {"duration": (1, 3, 5), "dependencies":
["Task H"], "condition": None},
    "Task J": {"duration": (4, 5, 8), "dependencies":
["Task I"], "condition": None}
    }
  #compute the average task durations
  baselineAverages = monteCarloAnalysis(tasks, n, "Monte
Carlo Risk Analysis")

  #Modify to have TD on one task (task C), which affects
one successor (task F)
  tasks = {
    "Task A": {"duration": (2, 5, 6), "dependencies":
[], "condition": None},
    "Task B": {"duration": (1, 2, 3), "dependencies":
["Task A"], "condition": None},
    "Task C": {"duration": (6, 10, 16), "dependencies":
["Task A"], "condition": None},
```

```
    "Task D": {"duration": (1, 5, 8), "dependencies":
["Task A"], "condition": None},
    "Task E": {"duration": (8, 15, 22), "dependencies":
["Task B"], "condition": None},
    "Task F": {"duration": (7, 10, 18), "dependencies":
["Task C"], "condition": {"dependent_on": ["Task C"],
"interest_amount": [(1,1.3,2)]}},
    "Task G": {"duration": (10, 15, 20), "dependencies":
["Task D"], "condition": None},
    "Task H": {"duration": (3, 10, 21), "dependencies":
["Task E", "Task F"], "condition": None},
    "Task I": {"duration": (1, 3, 5), "dependencies":
["Task H"], "condition": None},
    "Task J": {"duration": (4, 5, 8), "dependencies":
["Task I"], "condition": None}
    }
  #compute the average task durations
  tdAverages = monteCarloAnalysis(tasks, n, "Monte Carlo
Risk Analysis with Technical Debt")

  #Modify to have Compounding TD - technical debt on Task
C affects tasks F, G, H, I, and J
  tasks = {
    "Task A": {"duration": (2, 5, 6), "dependencies":
[], "condition": None},
    "Task B": {"duration": (1, 2, 3), "dependencies":
["Task A"], "condition": None},
    "Task C": {"duration": (6, 10, 16), "dependencies":
["Task A"], "condition": None},
    "Task D": {"duration": (1, 5, 8), "dependencies":
["Task A"], "condition": None},
    "Task E": {"duration": (8, 15, 22), "dependencies":
["Task B"], "condition": None},
    "Task F": {"duration": (7, 10, 18), "dependencies":
["Task C"], "condition": {"dependent_on": ["Task C"],
"interest_amount": [(1,1.3,2)]}},
    "Task G": {"duration": (10, 15, 20), "dependencies":
["Task D"], "condition": None},
    "Task H": {"duration": (3, 10, 21), "dependencies":
["Task E", "Task F"], "condition": {"dependent_on":
["Task C"], "interest_amount": [(1,1.3,2)]}},
    "Task I": {"duration": (1, 3, 5), "dependencies":
["Task H"], "condition": {"dependent_on": ["Task C"],
"interest_amount": [(1,1.3,2)]}},
    "Task J": {"duration": (4, 5, 8), "dependencies":
["Task I"], "condition": {"dependent_on": ["Task C"],
"interest_amount": [(1,1.3,2)]}}
    }
```

```
  #compute the average task durations
  compoundingAverages = monteCarloAnalysis(tasks, n,
"Monte Carlo Risk Analysis with Compounding Technical
Debt")

  #set up and plot the results
  width = 0.25
  multiplier = 0
  taskIDs = tasks.keys()
  taskValues = {}
  lbls = []
  taskValues["Baseline"] = baselineAverages
  taskValues["Technical Debt"] = tdAverages
  taskValues["Compounding Technical Debt"] = compound-
ingAverages
  x = np.arange(len(taskIDs))
  fig, ax = plt.subplots()
  # Set bar width and positions
  bar_width = 0.25
  # Create bar chart
  plt.bar(x-bar_width/2,            baselineAverages,
width=bar_width,    label='Baseline',    color='white',
edgecolor='black')
  plt.bar(x + bar_width/2, tdAverages, width=bar_width,
label='Technical Debt', color='gray', edgecolor='black')
  plt.bar(x   +   3*bar_width/2,   compoundingAverages,
width=bar_width,  label='Compounding  Technical  Debt',
color='black', edgecolor='black')
  # Add labels and title
  plt.xlabel('Task')
  plt.ylabel('Duration (days)')
  plt.title('Comparison of Task Durations due to Com-
pounding Technical Debt on Task C')
  plt.xticks(x + bar_width / 2, taskIDs)
  plt.legend()
  # Show chart
  plt.tight_layout()
  plt.show()
```

Appendix B
Example Technical Debt Register

Abstract An example technical debt register is provided in the context of a notional satellite development program. A narrative discussion of the technical compromises made during development accompanies tabular entries in the technical debt register. It is shown how the technical debt increased the duration of several tasks, leading to the need to make technical compromises on later tasks to meet the fixed schedule. This compounding technical debt is quantified and results in system with a high technical debt load at deployment.

Keywords Technical debt register

An example technical debt register is provided in Tables B.1, B.2, and B.3. Each project should tailor the fields to provide the information required and to fit the tools that they have. While the register below is in a tabular format, workflow management tools, such as Jira, could be used instead if they fit better with a project's current processes.

This example technical debt register was developed by considering the development of a notional communications satellite. This satellite is being launched in twelve months as a ride-share mission, which means that the satellite developer does not control the launch schedule and the satellite is subject to strict mass limits. If the satellite is late, then it will have to forfeit the original payment for the launch services and will have to wait for another launch opportunity, which will occur at an unknown time and cost. These constraints will be sources of stakeholder pressure on the system developer. The communications system on the satellite relies on a single antenna and must support multiple user terminals with different ground antenna characteristics. Of note with satellite systems is that the hardware cannot be changed after the satellite is launched, and therefore any performance improvements can only be obtained through software updates. Let's walk through a potential development cycle for this satellite and the decisions and compromises that might be made. Within this narrative, we will identify technical debt items with superscripts, which correspond to the entries in the technical debt register in Tables B.1, B.2 and B.3.

H. Kleinwaks, *Managing Technical Debt in Systems Engineering*,
https://doi.org/10.1007/978-3-032-30360-8

The development of the satellite starts with the identification of the stakeholder needs. Due to the short timeframe, the stakeholders identify only one ground-based terminal with which the satellite must communicate.[1] After the elicitation of stakeholder needs, the system requirements are developed. Initial modeling and simulation determine the required performance of the antenna for the satellite; however, this modeling and simulation is low-fidelity and relies on the experience of the satellite's chief engineer.[2] The belief that an existing antenna will work results in limited decomposition of performance requirements to the antenna subsystem. Additionally, even though the system mass is a highly constrained parameter based on the launch vehicle requirements, a mass budget is not flowed down to the antenna subsystem.[3]

With the requirements defined, the satellite moves into the preliminary design phase. During this phase, the initial models of the system performance are revisited, and it turns out that the chief engineer's instincts were not correct across all use cases. The system, as initially designed, does not close the link to the example terminal across all required locations at all atmospheric conditions, and the design needs to be reworked. The design problem is solved by increasing the size of the antenna to increase its gain.

The increase in antenna size causes an increase in the overall system mass, exceeding the launch vehicle allowance. Therefore, a technical trade is conducted, and it is decided that the only way to reduce the mass is to decrease the number of batteries, which reduces the duty cycle of the antenna.[4] During critical design, the guidance and navigation system of the satellite needs to be redesigned due to the change in the mass properties of the satellite, further delaying its development.

The satellite design is now behind schedule, and therefore it is decided to reduce the number of tests performed before the launch of the satellite.[5] Documentation is delayed where possible, and verification methods in the RVTM are changed from test to analysis.[6]

The satellite meets its launch date and is successfully inserted into its orbit by the launch vehicle. The satellite experiences a longer test and checkout time due to the limited testing that was performed. Many tests need to be repeated on-orbit, and the defect repair time is significantly increased due to slow software upload times and infrequent direct communication opportunities. Radiation events occur frequently, causing a reset of the satellite and resulting in the inability to consistently deliver service to the users. The limited duty cycle further reduces the service delivery. The users of the system are unsatisfied with the system, and it fails overall system validation.

This example shows how technical debt can build up within the system development process and lead to the failure of the system if it is not managed. If the satellite developer were keeping a technical debt register, then it would contain entries like those in the following tables. Actively managing this technical debt may have led to different decisions in the development process and to a more successful system. The first set of fields in the register (Table B.1) are the descriptive fields—those that provide information about the technical debt item. These fields should provide enough information to understand what the technical debt item is and how it affects the system. The next set of fields in the register (Table B.2) are the management fields.

These fields provide information about the current state of the technical debt item, including the progress toward mitigating its impact. Finally, we have the measurement fields (Table B.3). These fields provide the quantification of the technical debt item. Note that not all of these fields will apply to each item. For example, the fee amount does not apply to interest-bearing debt.

These tables show a snapshot of the technical debt register as it would have appeared at the time the satellite was launched and before any of the on-orbit issues were revealed.

What does this example technical debt register tell us about the progression of technical debt in a system development? Let's walk through the different items. First, we have the technical debt items that were repaid (items 1–3). Item 1, the under specification of end user systems, was repaid before the fee was realized. There were no taxes since no workarounds had to be undone, but principal (2 weeks) had to be paid to ensure that a system that would satisfy its users was being designed. Item 2, the failure to execute models of the proper fidelity, had to pay the interest (6 weeks) of redesigning the antenna in addition to the principal (4 weeks) of executing the correct models. This technical debt item produced a 10-week delay in the project schedule. Finally, we have Item 3, the failure to flow a mass budget requirement to the antenna vendor. This simple, unintentional oversight has significant ramifications for the project. If the antenna redesign associated with technical debt Item 2 comes in over mass, then additional work would be required to work around the new antenna mass, such as redesigning guidance algorithms. This simple source of technical debt can cost up to 6 weeks of delays, as was realized in this example.

As of the launch time of the satellite, Item 4 is accepted—the duty cycle of the vehicle is reduced based on technical trades on mass allowance, which results in technical fees and the reduction in performance. Items 5 and 6 are being mitigated—the impact of the limited testing has yet to be realized but is likely to cause a delay in the delivery of capability to the user. These are cases where the principal may never be able to be repaid, unless additional ground test assets are acquired. The acquisition of this ground test asset, which removes the workaround of testing in space, represents the taxes on these technical debt items. These technical debt items also carry technical fees—by not fully testing on the ground, there is a chance that the system will underperform and not be able to be corrected.

Figure B.1 shows the planned (black) and actual (gray) schedule for this notional satellite development program, based on the technical debt items identified in the technical debt register. Looking at this view, it is easy to see how schedule pressure could build on the system developer, based on the early schedule delays. The developer is starting the test phase with only a few weeks left to make the launch date, which encourages the compromise of delaying many tests and introduces delays to operations and the potential for additional technical fees.

The technical debt register also allows tracking of the system technical debt magnitude, if measurements are included in the register. Using the sequence of events outlined above, we can develop measurements of the system technical debt for this satellite development program, as shown in Fig. B.2. There is an initial accumulation of technical debt balance and interest due to the first three technical debt items, which

Table B.1 Descriptive fields of the technical debt register

ID	Descriptive fields			
	Technical debt statement	Description	Type	Artifact(s)
1	To preserve schedule, the satellite is only required to be compatible with an exemplar terminal, resulting in the potential for incompatibilities with other user terminals and a reduction in user satisfaction	The satellite is advertised as being compatible with multiple ground terminals. Since only one terminal is specified in the requirements, then there is the risk that incompatibilities will existing with other terminals. The lack of specification of atmospheric and environmental conditions also increases the likelihood of insufficient performance	Domain	Statement of needs and system CONOPs
2	To maintain schedule, high-fidelity modeling and simulation was not performed, resulting in the potential for underestimation of system performance	Relying on the chief engineer's intuition, only low-fidelity modeling and simulation was performed, which does not address the performance of the antenna across all potential operating environments. The antenna design may be insufficient across these environments and these insufficiencies may not be found until the antenna undergoes tests, which will result in the need for workarounds or acceptance of performance limitations	Modeling and Simulation	System models
3	Unintentionally, a mass budget was not provided to the antenna subsystem provider, resulting in the potential for overweight antennas	Without a mass budget, the antenna subsystem may come in overweight, which will require workarounds elsewhere in the satellite design to reduce mass. Additional work may be required to account for the increased antenna mass, such as working new guidance algorithms and adjustment of attitude control devices such as reaction wheels	Requirements	Antenna subsystem requirements documents

(continued)

Table B.1 (continued)

ID	Descriptive fields			
	Technical debt statement	Description	Type	Artifact(s)
4	To meet the launch mass and user connection requirements, the number of satellite batteries was reduced, resulting in the potential for decreased contact time with the users	To compensate for the increased antenna mass, a battery was removed. While this technical trade allows the system to meet the mass constraints of the launch vehicle, it also results in a reduced duty cycle for the satellite. The reduced duty cycle limits how frequently the antenna can be used and therefore restricts the number of users that can be serviced	Design	Spacecraft power system
5	To meet the launch schedule, the number of tests performed on the satellite are reduced, resulting in the potential for increased anomalies and longer resolution timelines	Since the project is running behind, tests are skipped to accelerate the schedule. These tests can be completed after launch and software updates loaded to the satellite. However, this process takes significantly longer since the satellite is not in constant contact with the ground and the upload data rates are limited. Only software corrective actions can be performed after launch	Test	Test plans
6	To meet the launch schedule, system verification is performed by analysis instead of test, resulting in the potential for increased anomalies and longer resolution timelines	To accelerate the schedule, verification is performed by analysis instead of tests. Analysis does not exercise the system, meaning that workmanship errors and other corner case errors may not be found prior to launch. These issues may be hardware issues and therefore may not be fixable after launch, or may require increased time to develop a software patch	Test	Verification products

are repaid by CDR (including both principal and interest). However, the decisions to forgo the testing due to the short timeline introduced much greater levels of principal and interest, which are still outstanding at the time of launch.

Table B.2 Management fields of the technical debt register

ID	Management fields		
	Status	Prioritization	Mitigation steps
1	Corrected (principal repaid)	Medium	• Identify all possible partner terminals • Identify requirements for atmospheric conditions • Calculate link budget to partner terminals across all atmospheric conditions • Update the list of possible terminals to only include those which can close the link
2	Corrected (principal repaid)	High	• Re-execute the models and simulations with higher fidelity parameters to determine link closure
3	Corrected (principal repaid)	High	• Flow a mass budget requirement to the manufacturer • Adjust guidance algorithms to account for higher mass allowance
4 ara>	Accepted	Medium	• Identify the duty cycle limitations due to the reduced battery count • Investigate new attitude algorithms to increase the charging time of the solar arrays • Investigate ability to vary the power profile of the antenna to trade data rate for duty cycle
5	Mitigating	High	• Ensure that all untested functions can be corrected via a software upload • Ensure that software upload paths are valid • Socialize an increased checkout timeline to the satellite users
6	Mitigating	High	• Ensure that all untested functions can be corrected via a software upload • Ensure that software upload paths are valid • Run additional analyses to examine corner cases • Design and execute specific on-orbit tests that compensate for skipped ground tests • Socialize an increased checkout timeline to the satellite users

Table B.3 Estimated values for measurement fields of the technical debt register

ID	Measurement fields					
	Principal	Taxes	Interest probability	Interest amount	Fee amount	Fee frequency
1	2 weeks (to properly determine the requirements)	N/A	N/A	N/A	80% reduction in number of terminals services: 4 out of 5 terminal types not considered	60% of users (most common terminal was the exemplar)
2	4 weeks (to execute the correct models)	N/A	50%	6 weeks (to redesign the antenna if errors found during test)	N/A	N/A
3	1 day (to flow the requirements)	N/A	75%	6 weeks (to redesign the guidance algorithms)	N/A	N/A
4	8 weeks (to reallocate the mass throughout the rest of the satellite)	N/A	N/A	N/A	25% reduction in duty cycle	25% of planned passes not usable
5	8 weeks (to execute the full test suite)	24 weeks + extra cost to develop a ground test asset	80%	16 weeks (to execute tests while on orbit and correct anomalies)	25% performance impact	10%
6	6 weeks (to execute the full test suite)	24 weeks + extra cost to develop a ground test asset	60%	12 weeks (to execute tests while on orbit and correct anomalies)	25% performance impact	10%

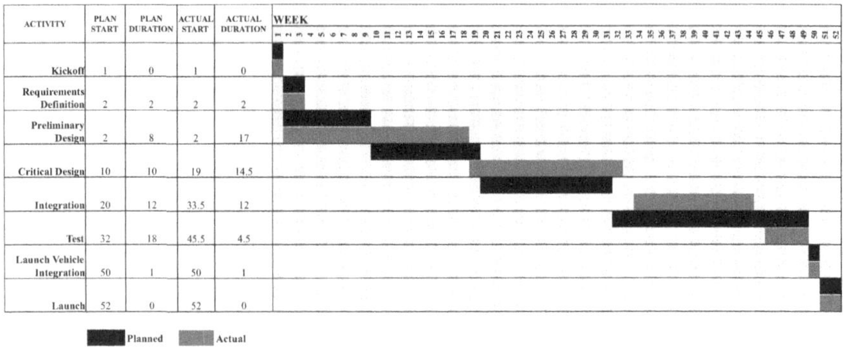

Fig. B.1 Satellite development program planned and actual schedule

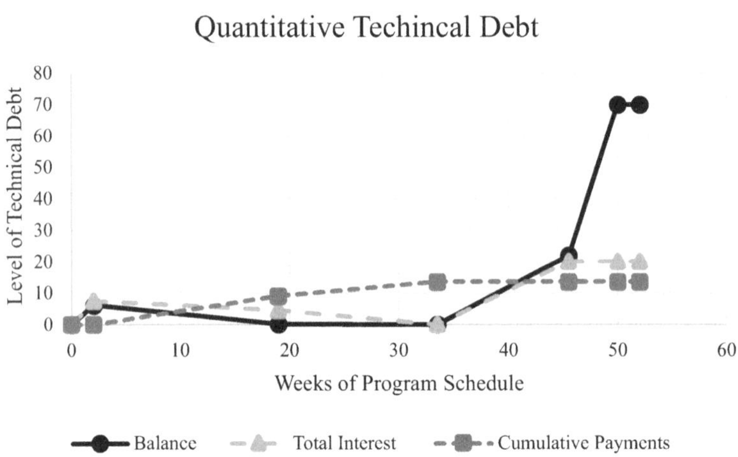

Fig. B.2 Quantitative measurements of technical debt

Appendix C
Technical Debt Review Board Agenda

Abstract An example agenda for a technical debt review board is provided. The agenda includes steps to monitor the progress of existing technical debt item repayment, to approve identified technical debt items, and to review areas of the system to identify new technical debt items. The technical debt review board also prioritizes the existing technical debt items.

Keywords Technical debt review board agenda

This appendix contains a sample agenda for a technical debt review board. Each project will need to tailor the review board to meet its needs. This exemplar agenda provides a starting point.

1. **Close completed technical debt items**: It is always nice to start meetings off on a positive note, and that can be accomplished by closing any technical debt items. Items can be closed when their mitigation plans have reduced the potential impact to acceptable levels or the principal and taxes have been repaid.
2. **Approve new technical debt items**: Any new technical debt items must be approved by the board. As stated in Sect. 11.1, new technical debt items can be brought forward by any member of the development team. The review board is responsible for ensuring that the proposed item meets the definition of technical debt and the descriptive standards required by the organization. Approval of the item subjects it to the management processes, including the execution of any mitigation plans. Any new risks created through acceptance of the technical debt item need to be documented for approval by the risk management board.
3. **Review existing technical debt items—mitigation status, measurements, and prioritization**: In the technical debt review board, it is essential to monitor the mitigation efforts, especially for the high-priority items. Reviewing this progress in the meeting, it reinforces the need for the team to be actively mitigating the impact of technical debt items. As each item is reviewed, updates should be made to the measured values (such as the interest probability or amount) and the prioritization. Both of these values may be affected by the progress of mitigation

H. Kleinwaks, *Managing Technical Debt in Systems Engineering*,
https://doi.org/10.1007/978-3-032-30360-8

steps and any changes to the system environment. It is recommended to start with the high-priority items to ensure that they are covered in the allotted time.

4. **Identify potential areas of technical debt and technical credits**: As the last agenda item, we recommend reviewing changes to the system since the last review board to identify any new areas where technical debt may have occurred or where any technical credits may be achieved. For example, if a new ICD update was released, the impacts of that update on the system and any corresponding changes should be reviewed for potential technical debt. Note that to keep the duration of the technical review board meeting short, it is usually best to identify areas in the system to examine for technical debt and to assign that work as an action to a board member, who can then report on the results at the next meeting. Potential technical credits should be identified as well—these are areas where the developer can invest additional time or money to increase the performance of the system. However, before doing so, the developer should bring the potential credit to the appropriate system review board.

At the conclusion of the technical debt review board, the technical debt register should be updated. As discussed in Sect. 7.2.4 and Sect. 11.1, making technical compromises and accepting technical debt may create new risks in the system. These risks need to be generated as an output of the technical debt review board and passed to the risk management board for incorporation in the risk management processes.

Index

H. Kleinwaks, *Managing Technical Debt in Systems Engineering*,
https://doi.org/10.1007/978-3-032-30360-8

V

W

The manufacturer's authorised representative in the EU is Springer Nature Customer Service Centre GmbH, Europaplatz 3, 69115 Heidelberg, Germany. If you have any concerns regarding our products, please contact ProductSafety@springernature.com

Printed and bound by CPI Group (UK) Ltd, Croydon, CR0 4YY
09/09/2026
02213715-0003